高等职业教育计算机类专业系列教材

Docker 容器技术应用

主　编　姜　雪　杨　忠　董长娥
副主编　张　楠　孙孟丽　赵文瀚

西安电子科技大学出版社

内 容 简 介

本书主要介绍Docker容器技术应用，书中将知识与技能相融合，以实现“教、学、做”一体化。全书共包含9个项目：创建Docker运行环境、安装Docker及管理镜像与容器、管理Docker数据与网络通信、创建Docker镜像、编排容器Docker Compose、部署和管理Harbor私有仓库、部署Docker安全、部署和管理Docker Swarm集群、部署和管理Kubernetes集群。各项目分为知识准备和任务实施两部分，前者主要介绍基础理论知识，后者侧重技能训练。各项目通过双创视角介绍云计算在我国各行各业的应用，并配有习题测试，以助力所学知识和技能的融会贯通。本书提供配套的安装软件、镜像和任务源代码等辅助学习资料，读者可登录西安电子科技大学出版社官方网站 (https://www.xduph.com) 自行下载。

本书可作为高等职业院校计算机相关专业的教材，也可作为计算机培训机构及计算机从业人员的参考用书。

图书在版编目 (CIP) 数据

Docker容器技术应用 / 姜雪，杨忠，董长娥主编 . -- 西安 : 西安电子科技大学出版社 , 2025. 2. -- ISBN 978-7-5606-7601-2

Ⅰ. TP316.85

中国国家版本馆CIP数据核字第2025JK5215号

策　　划　黄薇谚

责任编辑　黄薇谚　明政珠

出版发行　西安电子科技大学出版社 (西安市太白南路2号)

电　　话　(029) 88202421　88201467　　邮　　编　710071

网　　址　www.xduph.com　　电子邮箱　xdupfxb001@163.com

经　　销　新华书店

印刷单位　广东虎彩云印刷有限公司

版　　次　2025年2月第1版　2025年2月第1次印刷

开　　本　787毫米 × 1092毫米　1/16　印　张　13

字　　数　307千字

定　　价　40.00元

ISBN 978-7-5606-7601-2

XDUP 7902001-1

*** 如有印装问题可调换 ***

FOREWORD

前　言

云计算、大数据和人工智能是当今世界的热门技术，它们相互关联，共同推动着科技的进步和发展。云计算的应用涉及制造、金融、医疗、教育、政务等多个领域，逐渐成为产业发展的重要支撑。随着我国云计算技术的迅猛发展，企业上云也越来越普遍。

容器技术在云计算应用中发挥着重要作用，其中Docker作为一个开源的容器化平台，能够将应用程序及其依赖环境打包至一个容器中，并在任何支持Docker的环境中运行。Docker容器之间既互相隔离又可以彼此通信。

本书包含9个项目，分别是：项目一“创建Docker运行环境”，介绍Linux、CentOS和虚拟机的基础知识，并完成安装虚拟机软件、创建和初始化虚拟机等任务；项目二“安装Docker及管理镜像与容器”，介绍Docker容器技术与计算机虚拟化技术等知识，并完成安装Docker、体验镜像和容器的基本操作等任务；项目三“管理Docker数据与网络通信”，介绍Docker的数据存储、数据卷与数据卷容器以及Docker网络通信等知识，并完成管理Docker数据、建立端口映射、实现容器互联和自定义网络等任务；项目四“创建Docker镜像”，介绍Docker镜像的结构与创建方法、Dockerfile等知识，并完成通过容器和Dockerfile创建镜像等任务；项目五“编排容器Docker Compose”，介绍编排容器和Docker Compose等知识，并完成安装Docker Compose和使用Docker Compose部署服务等任务；项目六“部署和管理Harbor私有仓库”，介绍公有仓库和Harbor私有仓库等知识，并完成部署和管理Harbor私有仓库等任务；项目七“部署Docker安全”，介绍Docker安全、Cgroup资源管理和限制机制以及Docker日志等知识，并完成设置容器的CPU使用率与CPU周期、限制CPU内核 / 内存 /Block IO和查看Docker日志等任务；项目八“部署和管理Docker Swarm集群”，介绍Docker Swarm的工作原理等基本知识，并完成配置Docker Swarm集群各节点的系统环境、部署和管理Docker Swarm集群等任务；项目九“部署和管理 Kubernetes 集群”，介绍 Kubernetes 的体系架构和相关概念等基础知识，并完成配置 Kubernetes 集群各节点的系统环境、部署和管理 Kubernetes 集群等任务。

本书融入“1 + X 云计算平台运维与开发职业技能等级证书(中级)”“华为 HCIA/HCIE-Cloud Computing”“世界职业院校技能大赛云计算赛项”等考核内容与要求，形成“岗课融合、课证融合、课赛融合”的课程体系。

本书由淄博职业学院姜雪、杨忠和董长娥担任主编，由山东润图信息技术有限公司张楠和淄博职业学院孙孟丽、赵文瀚担任副主编，在此对参与本书编写的所有团队成员表示衷心感谢！

由于编者水平有限，书中难免存在疏漏和不妥之处，敬请广大读者批评指正。如对本书有任何问题，读者可发送邮件至 913617153@qq.com。

编　者

2024 年 8 月

CONTENTS

目　录

项目一
创建 Docker 运行环境

(1) 了解 Linux 的概念。
(2) 了解 CentOS 操作系统。
(3) 认识虚拟机。
(4) 掌握安装 VMware Workstation 虚拟机软件的方法。
(5) 掌握创建虚拟机的方法。
(6) 掌握配置虚拟机网络的方法。
(7) 掌握虚拟机连接远程管理工具的方法。
(8) 掌握初始化虚拟机的方法。

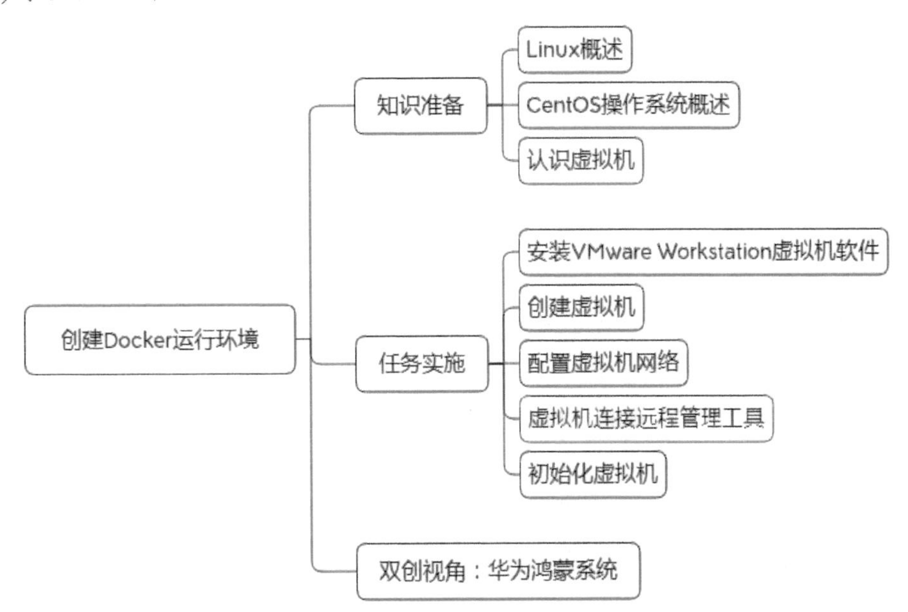

1.1 知识准备

1.1.1 Linux 概述

操作系统 (Operating System，OS)，是管理计算机硬件与软件资源的计算机程序。操作系统需要处理如管理与配置内存、决定系统资源供需的优先次序、控制输入设备与输出设备、操作网络与管理文件系统等基本事务，同时操作系统也提供了用户与系统进行交互操作的界面。常见的操作系统有 DOS 操作系统、Windows 操作系统、macOS 操作系统、UNIX 操作系统、Linux 操作系统等。

DOS(Disk Operating System，磁盘操作系统) 是 1979 年由微软公司为 IBM 个人计算机开发的 MS-DOS，是一个单用户单任务操作系统。DOS 一般都是黑底白色文字的界面，可直接操作和管理硬盘中的文件。从 1980 年到 1995 年，DOS 在 IBM PC 兼容机市场中占有举足轻重的地位。后来 DOS 的概念也包括了其他公司生产的与 MS-DOS 兼容的系统，如 PC-DOS、DR-DOS 等。DOS 操作系统的优点是使用一些接近于自然语言或其缩写的命令就可以轻松地完成绝大多数日常操作，系统占用空间小，对硬件配置要求低，能有效地管理、调度、运行个人计算机各种软件和硬件资源；缺点是缺乏图形界面，命令多，难度大，支持的硬件 / 软件少等。

Windows 操作系统也是由微软公司开发的，从 20 世纪 80 年代的 Windows 1.0 到现在的 Windows 10，Windows 操作系统在个人计算机中占据着绝对主流的地位，全球市场占有率高达 85% 左右，典型的代表有 Windows 98、Windows XP、Windows 7、Windows 10。Windows 系统采用图形化界面，是多用户、多任务操作系统，且应用程序众多；但是软件兼容性差，不开源，安全性较差，病毒多，移动端的支持较差等。

macOS 操作系统是由苹果公司开发的一款高性能计算机操作系统，在图形设计、影视制作等领域有着自己独特的优势，所以专业设计公司大多使用的是安装 macOS 系统的苹果电脑。macOS 系统兼容性好，安全性高，系统稳定，操作便捷；但是外接配件需转接口，软件限制大，内存和硬盘升级难等。

UNIX 操作系统是一款付费系统，诞生于 20 世纪 70 年代，其安全性非常高，大多用在企业级别的服务器上，如银行、电信公司等的服务器，很多公司的研发系统也使用了 UNIX 系统平台。UNIX 系统的可靠性高，可伸缩性强，开放性良好，网络功能强大，数据库支持能力强等；但 UNIX 的系统调用接口复杂，缺乏集成度高且功能全的程序，内核不够灵活等。

Linux 是一款免费使用和自由传播的类 UNIX 操作系统。早期的 Linux 系统跟微软的 DOS 系统一样，采用命令行字符操作模式，因为没有图形界面，所以只有一些专业领域 (如网站运维) 的人员才会用到。后来，Linux 系统也慢慢推出了图形模式，之后才开始在个人计算机上使用，如 deepin 深度操作系统、CentOS 等。Linux 系统和 Windows 系统的比较见表 1-1 所示。

表 1-1　Linux 系统和 Windows 系统的比较

操作系统	占用硬件资源	服务器操作系统性能	稳定性	安全性	远程管理	资源利用率
Windows 系统	较多	较低	较低	较低	效率较低	效率较低
Linux 系统	较少	较高	较高，可保持开机几年	较高	高效	高效

Linux 内核由 Linus Benedict Torvalds 于 1991 年初次开发，因此 Linus 被称为 Linux 之父。Linux 主要受到 Minix 和 UNIX 思想的启发，是一个基于 POSIX 的多用户、多任务、支持多线程和多 CPU 的操作系统。POSIX(Portable Operating System Interface，可移植操作系统接口) 是 IEEE(Institute of Electrical and Electronics Engineers，美国电气及电子工程师学会) 为要在各种 UNIX 操作系统上运行软件而定义 API(Application Program Interface，应用程序接口) 的一系列互相关联的标准的总称，X 表明其是对 UNIX API 的传承。

Linux 有很多不同的发行版，如基于社区开发的 CentOS、Debian、Ubuntu、Arch Linux 等，基于商业开发的 Red Hat Enterprise Linux、SUSE、Oracle Linux 等。Linux 操作系统有以下特点：

(1) 免费的自由软件。Linux 系统是通过公共许可协议 (General Public License，GPL) 的自由软件，它开放并免费提供源代码，开发者可以根据自身需要自由修改、复制和发布程序的源代码。因此，用户可以从互联网上很方便地免费下载并使用 Linux 操作系统，不需要考虑版权问题。

(2) 良好的硬件平台可移植性。操作系统从一个硬件平台转移到另一个硬件平台时，只需改变底层的少量代码，无须改变自身的运行方式。Linux 系统几乎能在所有主流 CPU 搭建的体系结构上运行，包括 Intel/AMD、HP-PA、MIPS、PowerPC、UltraSPARC 和 Alpha 等，其伸缩性超过了所有其他类型的操作系统。

(3) 完全符合 POSIX 标准。这主要是为了增强不同操作系统之间的兼容性和应用程序的可移植性。

(4) 良好的图形用户界面。Linux 系统具有类似 Windows 操作系统的图形界面，即 X-Window。X-Window 是一种起源于 Linux 操作系统的标准图形界面，它可以为用户提供一种具有多种窗口管理功能的对象集成环境。

(5) 强大的网络功能。由于 Linux 系统是依靠互联网平台迅速发展起来的，所以也具有强大的网络功能。它在内核中实现了 TCP/TP 协议簇，提供了对 TCP/TP 协议簇的支持。同时，它还可以支持其他各种类型的通信协议，如 IPX/SPX、Apple Talk、PP、SLP 和 ATM 等。

(6) 丰富的应用程序和开发工具。由于 Linux 系统具有良好的可移植性，目前绝大部分在 Linux 系统下可使用的流行软件都已经被移植到 Linux 系统中。另外，由于 Linux 得到了 IBM、Intel、Oracle 及 Sybase 等知名公司的支持，这些公司的诸多软件也都移植到了 Linux 系统中，因此，Linux 获得了越来越多的应用程序和应用开发工具。

(7) 良好的安全性和稳定性。Linux 系统采取了多种安全措施，如任务保护机制、审计跟踪、核心校验、访问授权等，为网络多用户环境中的用户提供了强大的安全保障。由于 Linux 系统的开放性，其具有良好的计算机病毒防御机制，因此 Linux 平台基本上不需要安装防病毒软件。另外，Linux 系统具有极强的稳定性，可以长时间稳定地运行。

1.1.2　CentOS 操作系统概述

CentOS(Community Enterprise Operating System，社区企业操作系统) 是一个基于 RHEL (Red Hat Enterprise Linux，红帽企业 Linux) 源代码的开源操作系统，旨在提供一个与红帽企业 Linux 兼容但不包含红帽企业支持的平台。CentOS 于 2004 年发布了首个版本，之后一直保持着新版本的更新迭代，在服务器行业中其因稳定性高、安全性好、软件包丰富等优点受到了广大用户的普遍欢迎。

2020 年 12 月 8 日，CentOS 项目宣布 CentOS 8 于 2021 年底结束使用，而 CentOS 7 则在 2024 年 6 月 30 日结束其生命周期，此后停止维护，不再针对 CentOS Linux 7 系统发布软件更新和补丁。这对使用 CentOS 的企业和个人产生了很大的影响，原先的系统可能面临安全风险、性能问题以及与其他软件的兼容性问题等。

我国操作系统经过数十年的探索和钻研，逐渐登上历史舞台，诸如银河麒麟、中标麒麟、统信 UOS、红旗 Linux 等操作系统逐步成熟，OpenEuler、OpenAnolis 等开源社区相继面世，大量国内服务器操作系统崭露头角、迅猛发展。

1.1.3　认识虚拟机

虚拟机是在实体计算机上通过软件模拟、具有完整硬件系统功能、运行在一个完全隔离环境中的完整计算机系统。根据用户需求，通常会模拟出一台或多台虚拟机，虚拟机使用宿主机的硬件资源，拥有真实计算机的绝大多数功能。在虚拟机中可以安装虚拟机软件所支持的操作系统，如 Linux 或 Windows，且与宿主机所使用的具体操作系统无关。

常见的虚拟机软件有 VirtualBox、VMware Workstation、Parallels Desktop、Hyper-V、QEMU 等。其中 VMware Workstation 是目前实现虚拟化程度最高、应用最广泛的虚拟化产品。

VMware Workstation 是一款功能强大的桌面虚拟计算机软件，它允许操作系统和应用程序在一台虚拟机内部运行，可以在运行于桌面上的多台虚拟机之间切换。虚拟机软件的挂起、恢复和退出等操作不会影响任何操作系统和宿主机的工作，也不会影响其他正在运行的应用程序。VMware Workstation 先进的技术与更好的灵活性超过了市面上其他的虚拟计算机软件，获得了更多用户的青睐。

VMware Workstation 主要具有以下特点：

(1) 支持 Windows、Linux 和 macOS 等多种操作系统；

(2) 可以创建桥接、NAT 和主机模式的网络配置以进行虚拟机通信；

(3) 虚拟机和宿主机之间能够方便地进行数据共享；

(4) 具有可视化图像操作功能；

(5) 具有加密数据传输、控制虚拟机访问和编辑主机文件等安全措施；

(6) 具有虚拟机快照功能，能够备份整个虚拟机的任何运行状态，当虚拟机出现问题时可以迅速还原；

(7) 具有虚拟机克隆功能，能够在很短的时间内复制出多台配置相同的虚拟机，方便搭建集群。

本书的任务实施将在具有 Windows 操作系统的计算机上，通过 VMware Workstation 虚拟机软件创建出的虚拟机来完成。

1.2 任务实施

1.2.1 安装 VMware Workstation 虚拟机软件

安装 VMware Workstation 虚拟机软件

1. 任务目标

掌握安装 VMware Workstation Pro16 虚拟机软件的方法。

2. 任务内容

(1) 下载 VMware Workstation Pro16 安装程序及注册码。

(2) 安装 VMware Workstation Pro16。

3. 完成任务所需的设备和软件

(1) 一台安装 Windows 10 操作系统的计算机。

(2) VMware Workstation Pro16 安装包及注册码。

4. 任务实施步骤

第一步，通过随书附带资源将 VMware16 下载到本地，如图 1-1 所示，然后双击 VMware 安装程序。

› 此电脑 › 本地磁盘 (D:) › jx › VMware16

名称	修改日期	类型	大小
VMware-workstation-full-16.1.2-179...	2023/12/29 9:39	应用程序	636,349 KB
注册码.txt	2022/1/18 17:30	文本文档	1 KB

图 1-1　将 VMware16 下载到本地

第二步，在 VMware Workstation Pro 安装向导界面单击“下一步”按钮，如图 1-2 所示。

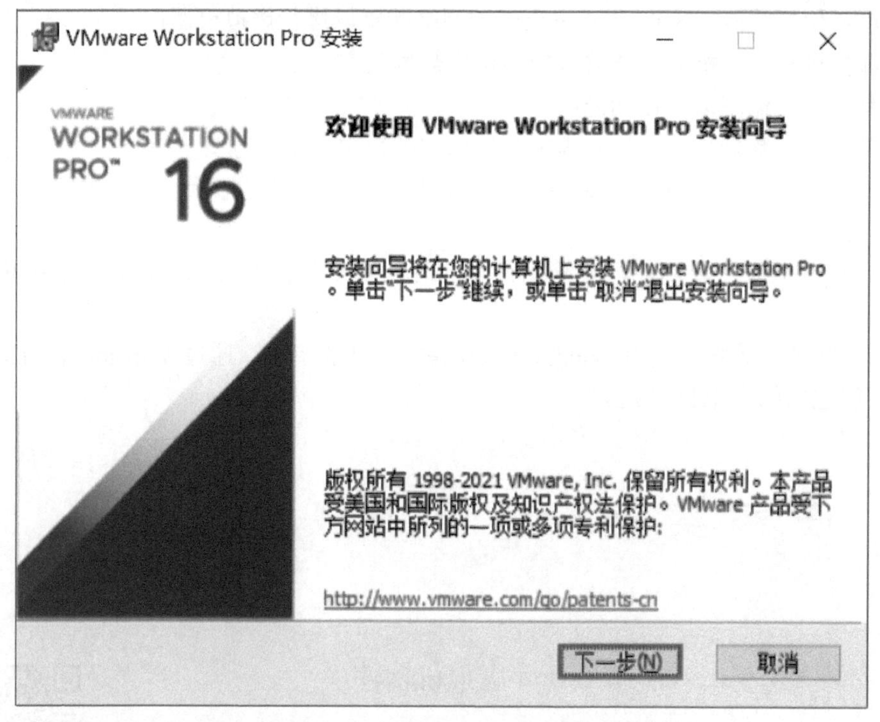

图 1-2　开始安装 VMware Workstation Pro16

第三步，接受许可协议，单击“下一步”按钮，如图 1-3 所示。

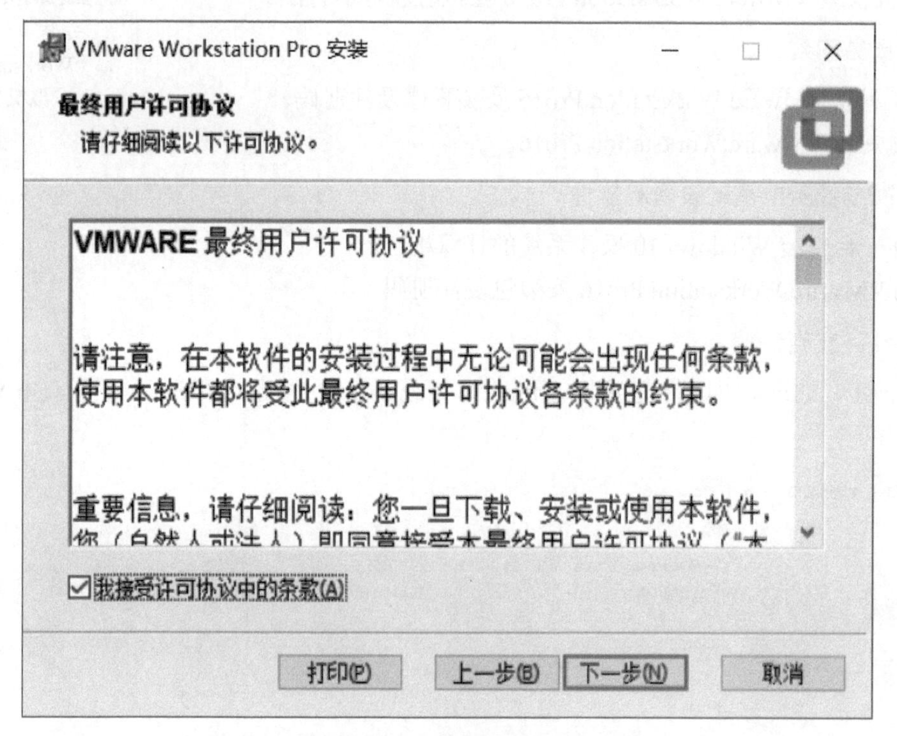

图 1-3　接受许可协议

第四步，更改软件安装位置，建议安装在除 C 盘以外的任意盘符下，单击“下一步”按钮，如图 1-4 所示。

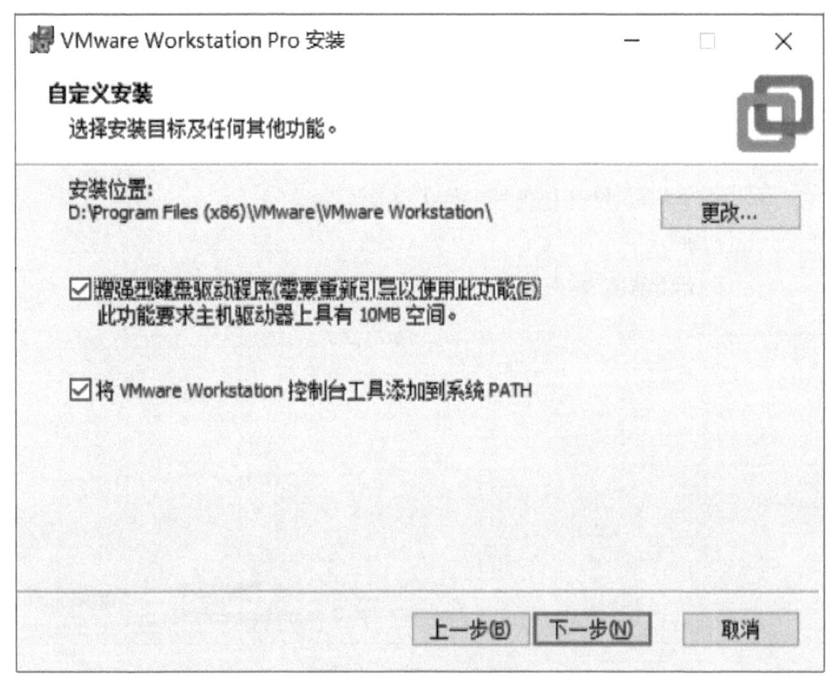

图 1-4 更改安装位置

第五步，在用户体验设置界面，为了避免干扰可不勾选两个复选框，直接单击“下一步”按钮，如图 1-5 所示。

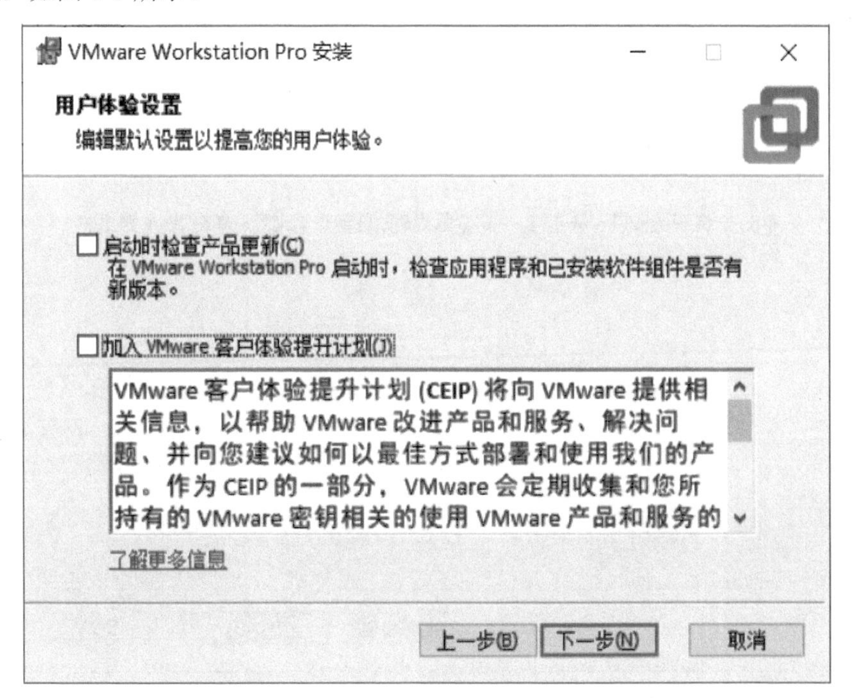

图 1-5 用户体验设置

第六步，默认快捷方式的设置，单击“下一步”按钮，如图1-6所示。

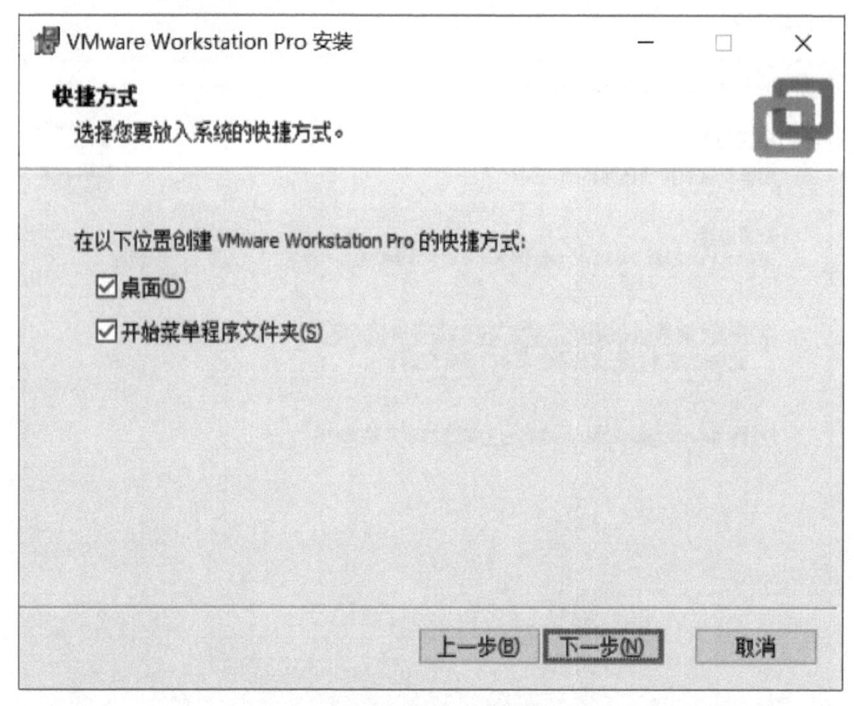

图1-6 快捷方式设置

第七步，在已准备好安装VMware Workstation Pro界面，单击“安装”按钮，如图1-7所示。

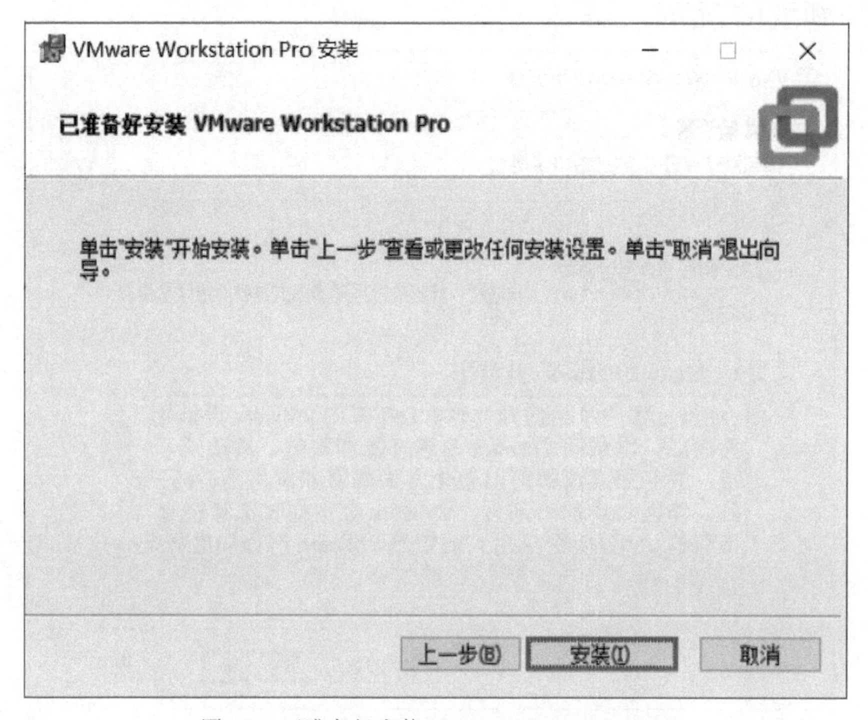

图1-7 已准备好安装VMware Workstation Pro

第八步，安装进行中，等待安装完成，如图 1-8 所示。

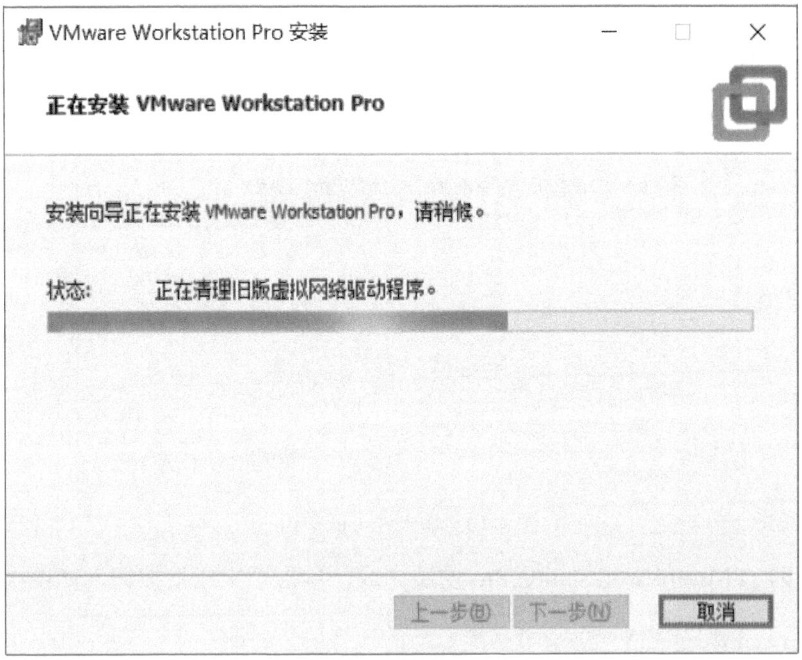

图 1-8 安装进行中

第九步，单击“许可证”按钮，如图 1-9 所示。

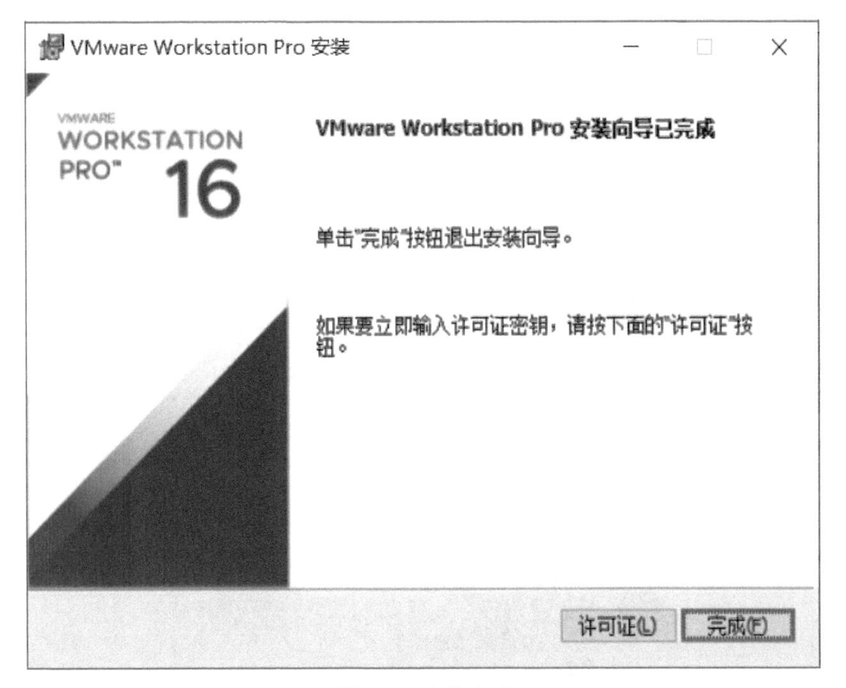

图 1-9 安装完成

第十步，从激活码中任意复制一个号码到文本框中，单击“输入”按钮，激活软件，如图 1-10 所示。

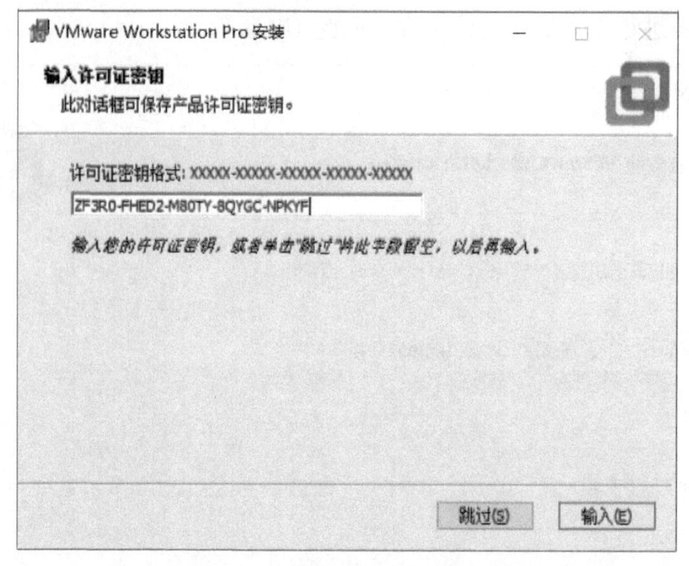

图 1-10　输入许可证密钥

第十一步，VMware Workstation Pro 安装完成，单击“完成”按钮，如图 1-11 所示。

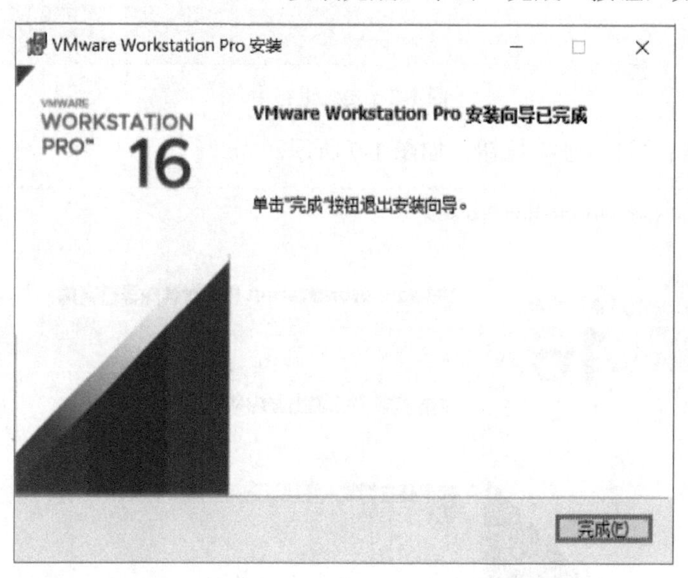

图 1-11　VMware Workstation Pro 安装完成

第十二步，重启系统，如图 1-12 所示。

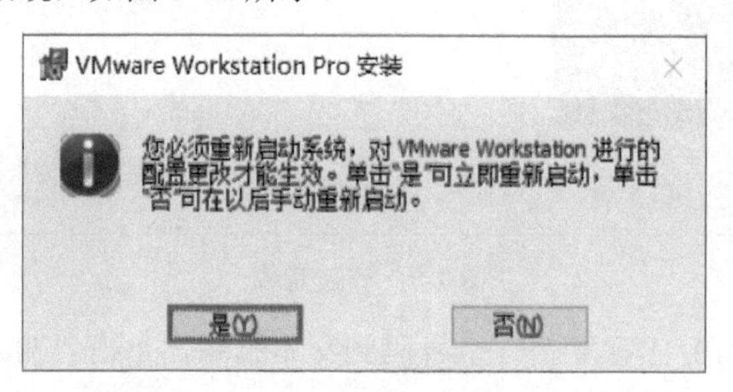

图 1-12　重启系统

第十三步，虚拟机软件 VMware Workstation 安装成功后，可以看到两个虚拟网卡 VMnet1 和 VMnet8 已经存在，如图 1-13 所示。

图 1-13　查看虚拟网卡 VMnet1 和 VMnet8

1.2.2　创建虚拟机

创建虚拟机

1. 任务目标

掌握创建安装 CentOS7.6 操作系统的虚拟机的方法。

2. 任务内容

(1) 下载 CentOS7 镜像。

(2) 创建虚拟机运行环境。

(3) 通过 VMware Workstation Pro 创建虚拟机。

3. 完成任务所需的设备和软件

(1) 一台安装 Windows 10 操作系统的计算机。

(2) VMware Workstation。

(3) CentOS7.6 镜像。

4. 任务实施步骤

第一步，通过地址 https://archive.kernel.org/centos-vault/7.6.1810/isos/x86_64/ 下载 CentOS7.6 镜像，或从附带资源中找到该镜像，保存到指定目录，如图 1-14 所示。

Index of /centos-vault/7.6.1810/isos/x86_64/

```
../
0_README.txt                                01-Dec-2018 13:21     2495
CentOS-7-x86_64-DVD-1810.iso                25-Nov-2018 23:55       4G
CentOS-7-x86_64-DVD-1810.torrent            03-Dec-2018 15:03      86K
CentOS-7-x86_64-Everything-1810.iso         26-Nov-2018 14:28      10G
CentOS-7-x86_64-Everything-1810.torrent     03-Dec-2018 15:03     101K
CentOS-7-x86_64-LiveGNOME-1810.iso          24-Nov-2018 17:41       1G
CentOS-7-x86_64-LiveGNOME-1810.torrent      03-Dec-2018 15:03      28K
CentOS-7-x86_64-LiveKDE-1810.iso            24-Nov-2018 17:53       2G
CentOS-7-x86_64-LiveKDE-1810.torrent        03-Dec-2018 15:03      37K
CentOS-7-x86_64-Minimal-1810.iso            25-Nov-2018 21:25     918M
CentOS-7-x86_64-Minimal-1810.torrent        03-Dec-2018 15:03      36K
CentOS-7-x86_64-NetInstall-1810.iso         25-Nov-2018 16:21     507M
CentOS-7-x86_64-NetInstall-1810.torrent     03-Dec-2018 15:03      20K
sha256sum.txt                               01-Dec-2018 13:16      598
sha256sum.txt.asc                           03-Dec-2018 14:50     1458
```

图 1-14　下载 CentOS7.6 镜像

第二步，打开 VMware Workstation 窗口，单击“创建新的虚拟机”打开新建虚拟机向导，如图 1-15 所示。

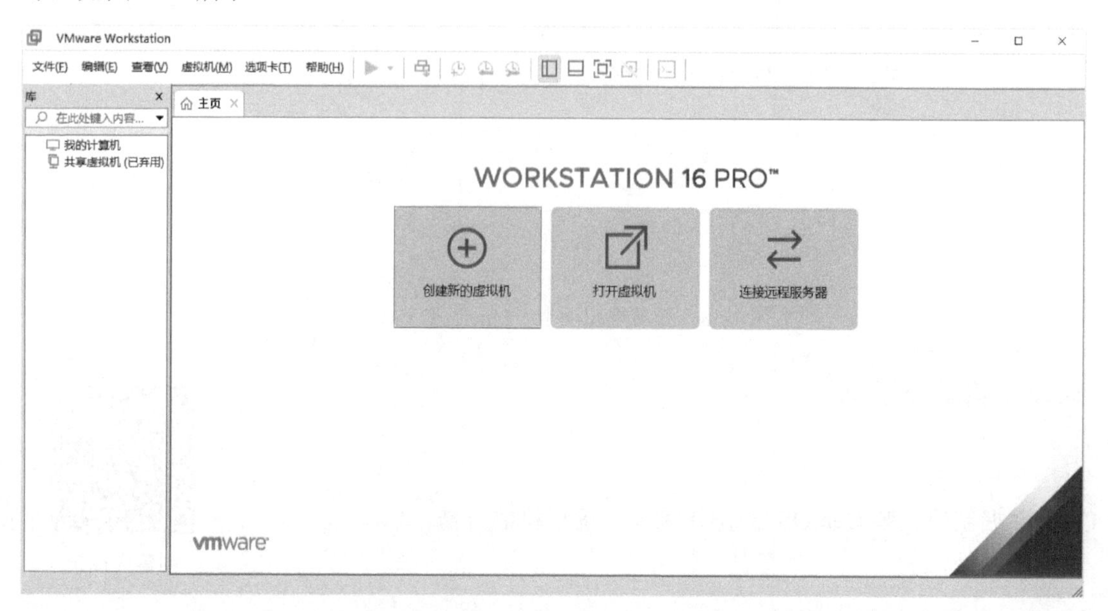

图 1-15　打开 VMware Workstation 窗口

第三步，选择“自定义”类型的配置，单击“下一步”按钮，如图 1-16 所示。

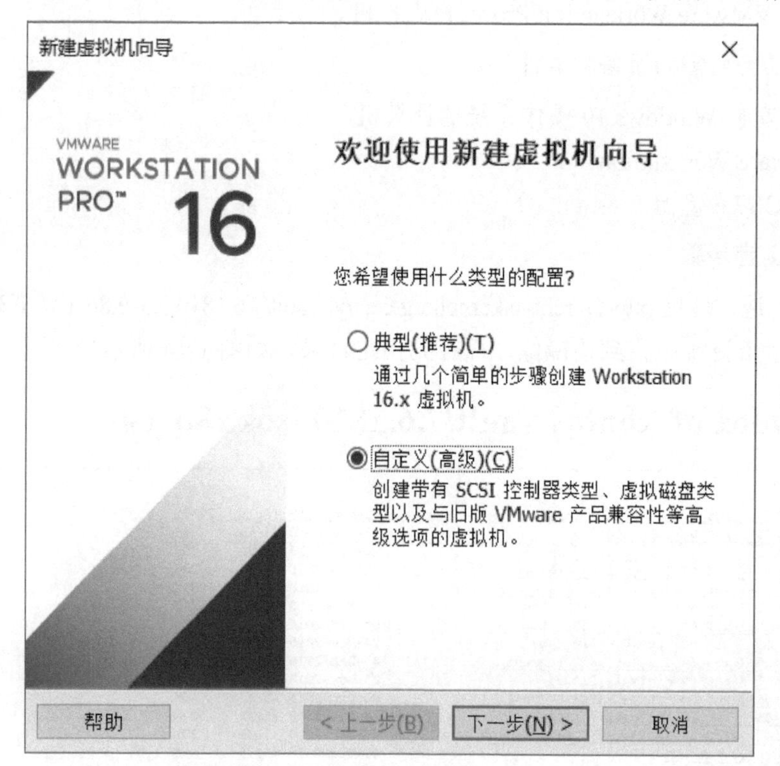

图 1-16　选择自定义类型的配置

第四步，虚拟机硬件兼容性选择默认即可，直接单击“下一步”按钮，如图 1-17 所示。

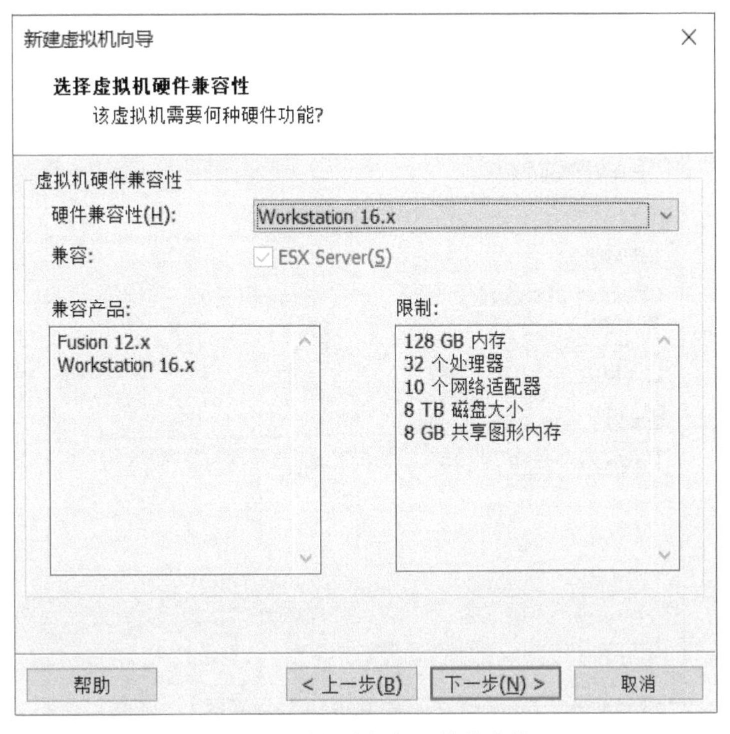

图 1-17　选择虚拟机硬件兼容性

第五步，安装客户机操作系统，选择“稍后安装操作系统”，单击“下一步”按钮，如图 1-18 所示。

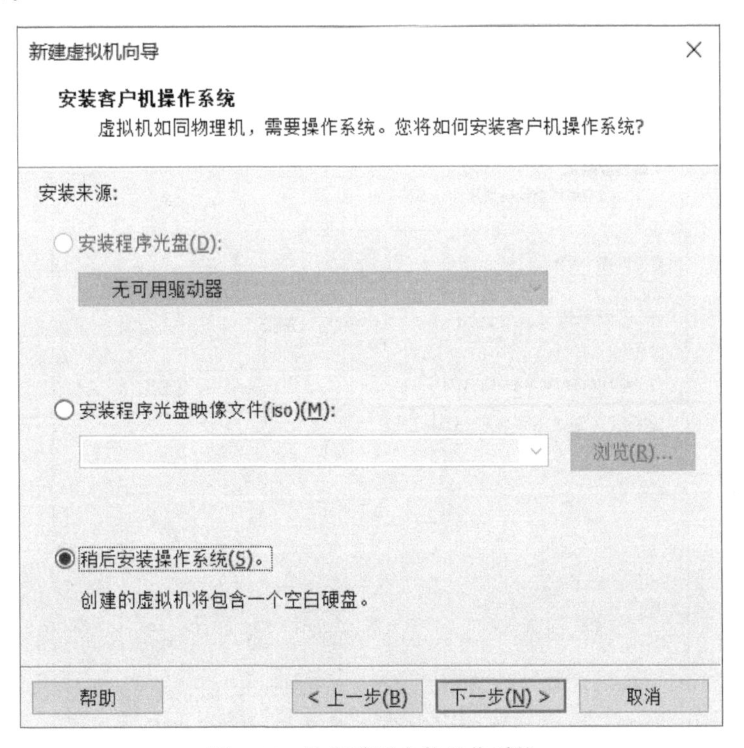

图 1-18　选择稍后安装操作系统

第六步，客户机操作系统选择“Linux”，版本选择“CentOS 7 64 位”，单击“下一步”按钮，如图 1-19 所示。

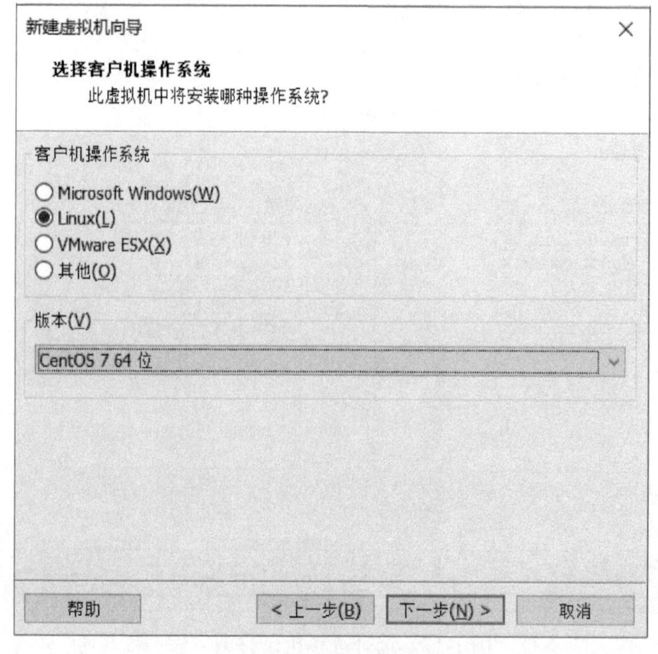

图 1-19　选择操作系统

第七步，设置虚拟机的名称和安装位置，单击“下一步”按钮，如图 1-20 所示。

注意：要为每台虚拟机配置一个文件夹，将虚拟机的所有文件放置其中，便于对多台虚拟机文件分别进行修改时不会产生混淆。

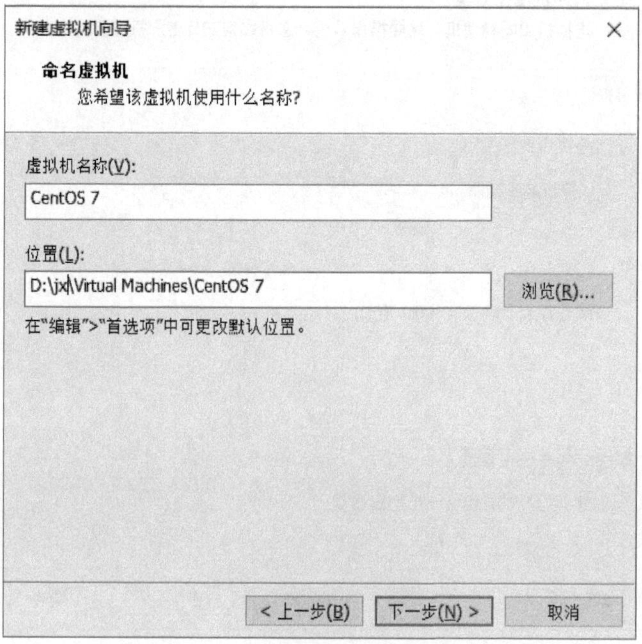

新建虚拟机向导
命名虚拟机
您希望该虚拟机使用什么名称?
虚拟机名称(V):
CentOS 7
位置(L):
D:\jx\Virtual Machines\CentOS 7
浏览(R)...
在“编辑”>“首选项”中可更改默认位置。
< 上一步(B)　下一步(N) >　取消

图 1-20　设置虚拟机的名称和安装位置

第八步，设置处理器的内核总数为 2，单击“下一步”按钮，如图 1-21 所示。

图 1-21　设置处理器的内核总数

第九步，设置虚拟机的内存为 2 GB，单击“下一步”按钮，如图 1-22 所示。

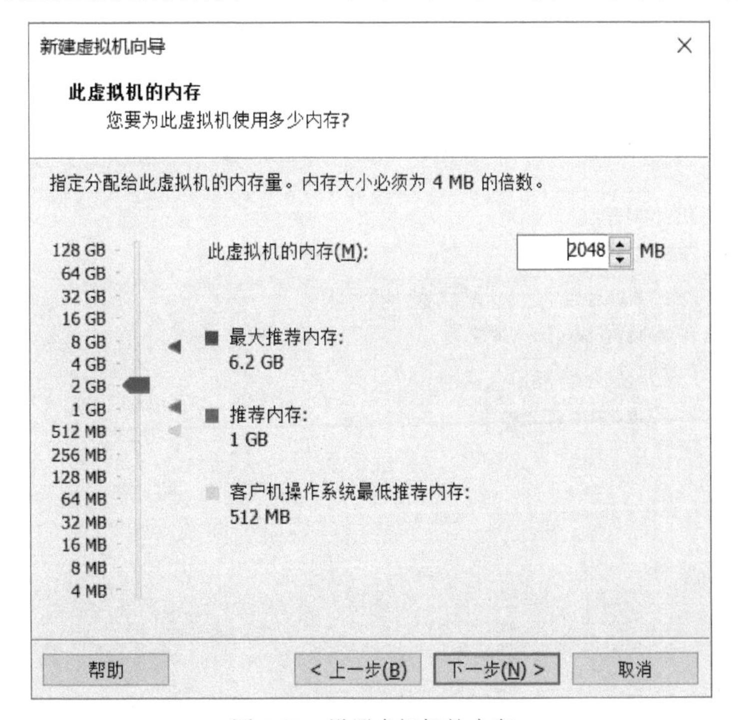

图 1-22　设置虚拟机的内存

第十步，网络类型选择默认“使用网络地址转换 (NAT)”，使得虚拟机和物理主机系统共享一个网络，单击“下一步”按钮，如图 1-23 所示。

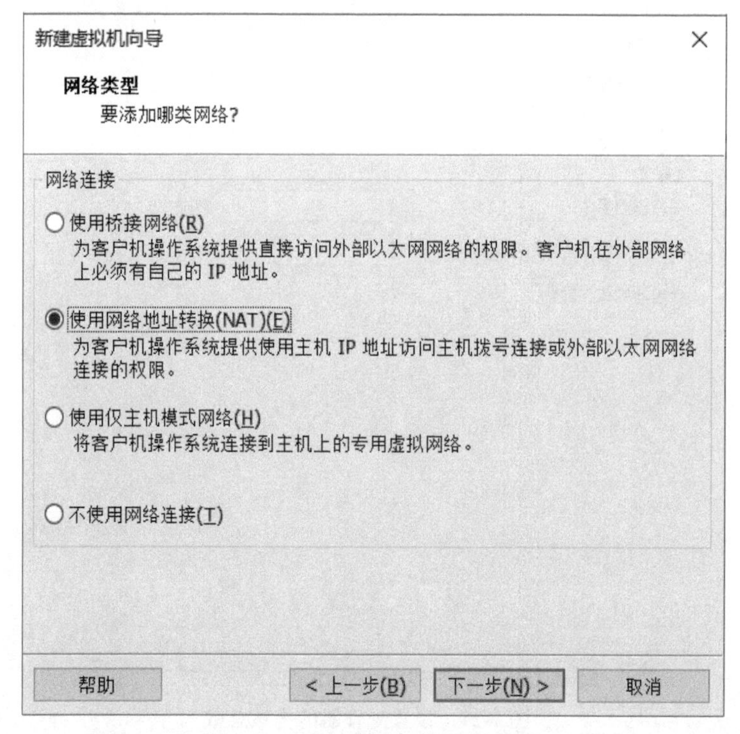

图 1-23　选择虚拟机的网络类型

第十一步，I/O 控制器类型选择默认设置，单击“下一步”按钮，如图 1-24 所示。

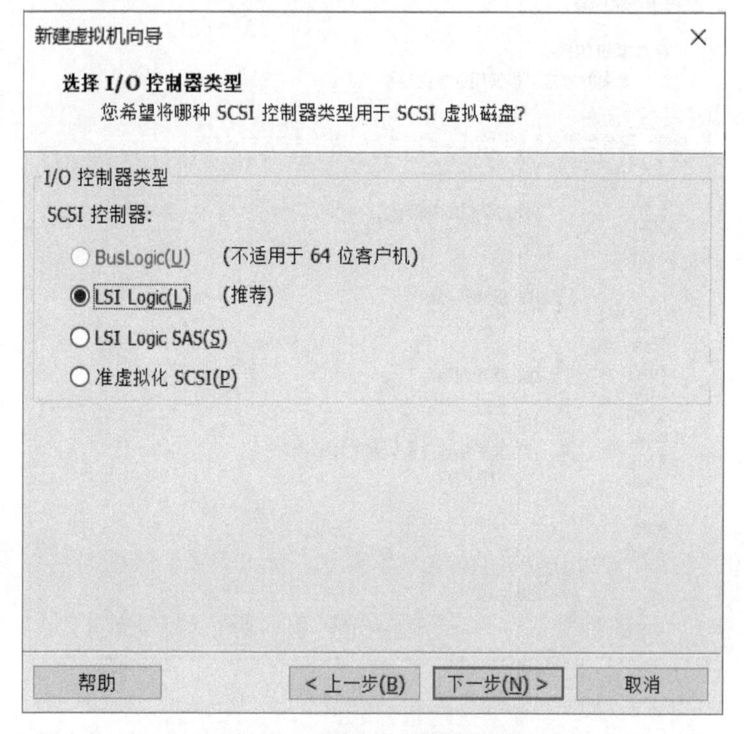

图 1-24　选择 I/O 控制器类型

第十二步，磁盘类型选择默认设置，单击“下一步”按钮，如图 1-25 所示。

新建虚拟机向导

选择磁盘类型

您要创建何种磁盘?

虚拟磁盘类型

IDE(I)

SCSI(S)　(推荐)

SATA(A)

NVMe(V)

帮助　< 上一步(B)　下一步(N) >　取消

图 1-25　选择磁盘类型

第十三步，磁盘选择默认设置，单击“下一步”按钮，如图 1-26 所示。

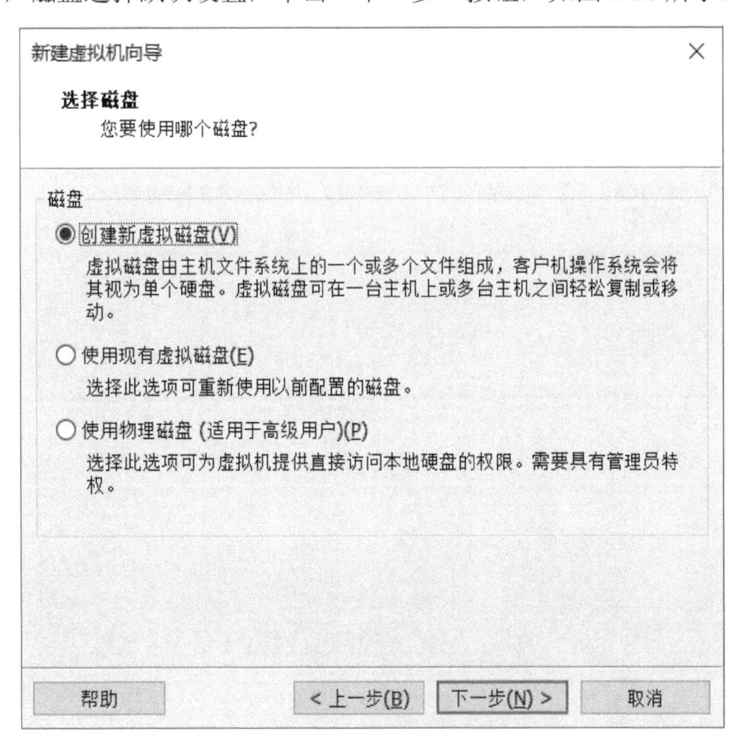

图 1-26　选择磁盘

第十四步，指定磁盘容量选择默认设置，单击“下一步”按钮，如图 1-27 所示。

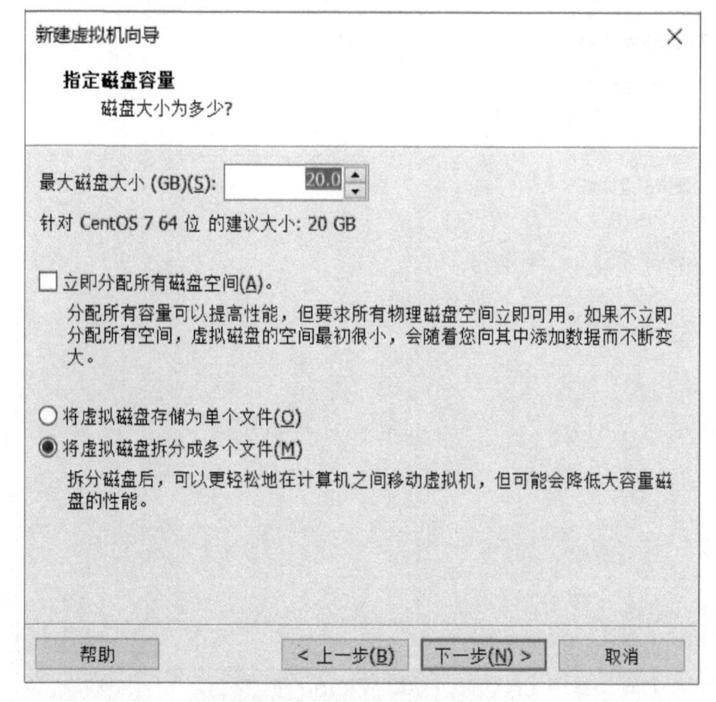

图 1-27　指定磁盘容量

第十五步，指定磁盘文件选择默认设置，单击“下一步”按钮，如图 1-28 所示。

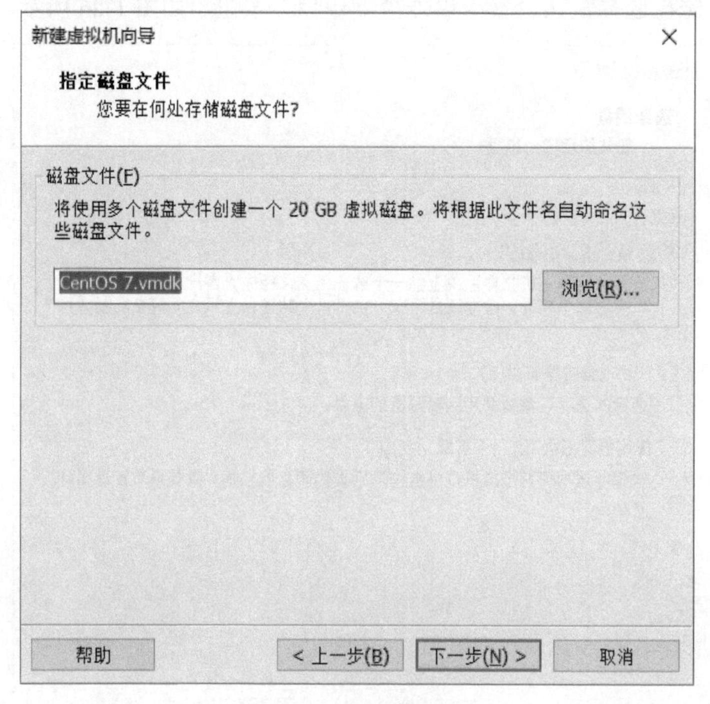

图 1-28　指定磁盘文件

第十六步，虚拟机运行环境已创建好，单击“完成”按钮，如图 1-29 所示。

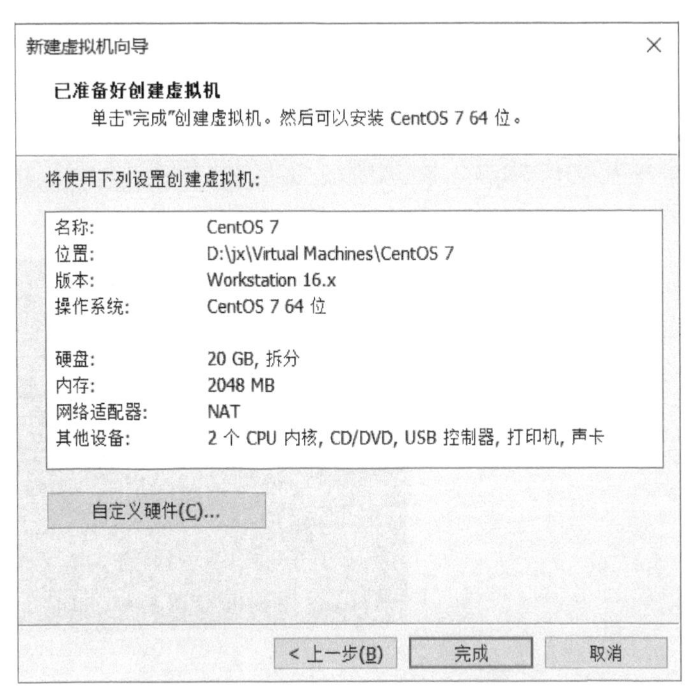

图 1-29　创建虚拟机运行环境

第十七步，在 VMware Workstation 窗口中，单击左侧“我的计算机”下的 CentOS7，在右侧 CentOS7 界面中单击“编辑虚拟机设置”，在打开的“虚拟机设置”对话框中单击“CD/DVD”，“设备状态”选择“启动时连接”，“连接”选择“使用 ISO 映像文件”，再单击“浏览”按钮打开本地 CentOS 镜像文件，然后单击对话框的“确定”按钮，如图 1-30 所示。

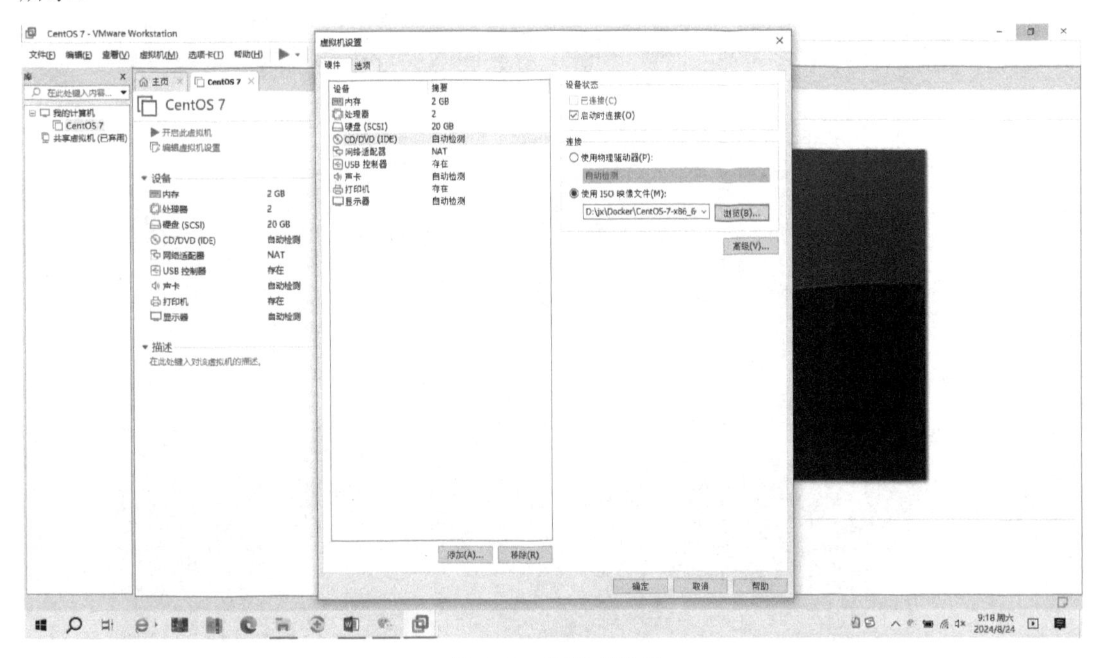

图 1-30　配置虚拟机

第十八步，在 VMware Workstation 窗口的 CentOS7 界面中，单击“开启此虚拟机”，如图 1-31 所示。

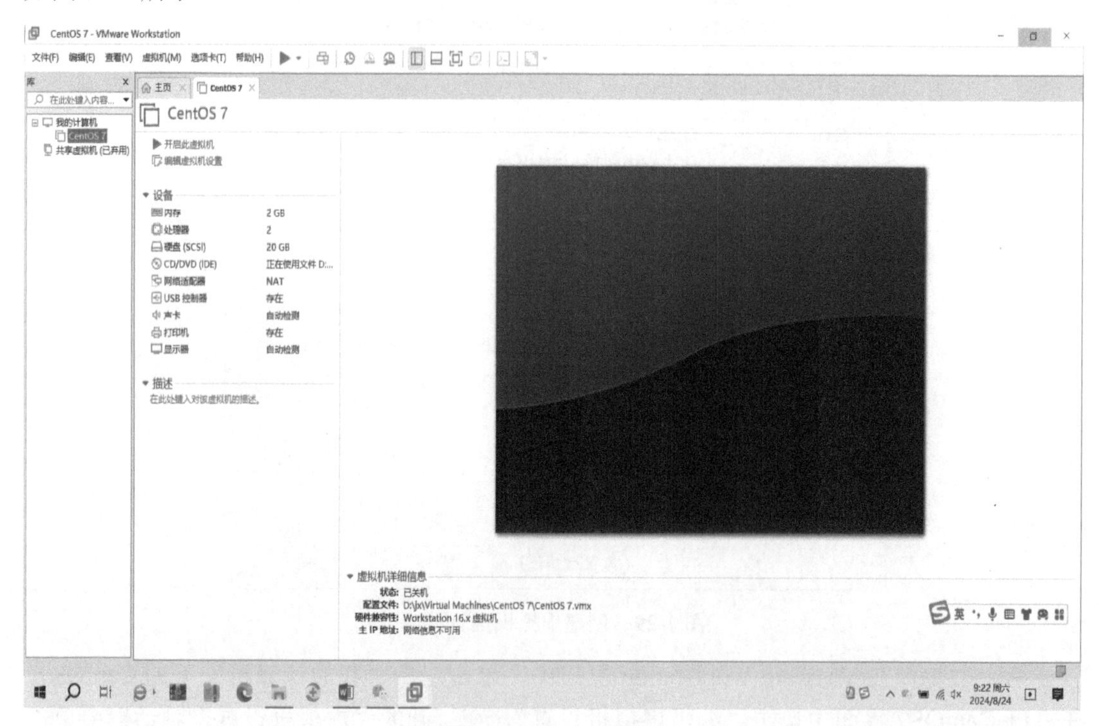

图 1-31　开启此虚拟机

第十九步，鼠标单击 CentOS7 虚拟机界面，用方向键“↑”选择“Install CentOS 7”并点击回车键，开始安装操作系统，如图 1-32 所示，安装过程如图 1-33 所示。

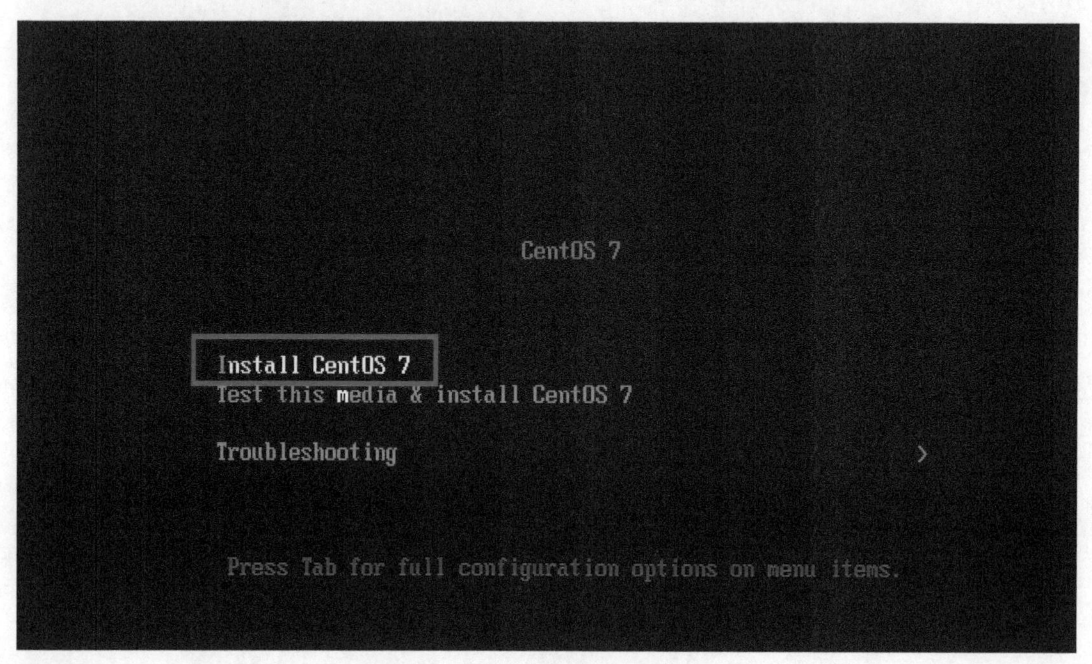

图 1-32　安装 CentOS 7 操作系统

```
[  OK  ] Listening on Delayed Shutdown Socket.
[  OK  ] Listening on udev Kernel Socket.
         Mounting Huge Pages File System...
[  OK  ] Created slice User and Session Slice.
[  OK  ] Listening on LVM2 metadata daemon socket.
[  OK  ] Reached target Paths.
         Mounting POSIX Message Queue File System...
         Starting Create list of required static device nodes for the current kernel...
[  OK  ] Reached target Slices.
[  OK  ] Started Remount Root and Kernel File Systems.
         Starting Configure read-only root support...
         Starting Rebuild Hardware Database...
[  OK  ] Mounted Temporary Directory.
[  OK  ] Mounted Huge Pages File System.
[  OK  ] Mounted Debug File System.
[  OK  ] Mounted POSIX Message Queue File System.
[  OK  ] Started Read and set NIS domainname from /etc/sysconfig/network.
[  OK  ] Started Create list of required static device nodes for the current kernel.
         Starting Create Static Device Nodes in /dev...
[  OK  ] Started Apply Kernel Variables.
[  OK  ] Started Journal Service.
         Starting Flush Journal to Persistent Storage...
[  OK  ] Started Configure read-only root support.
         Starting Load/Save Random Seed...
[  OK  ] Started Create Static Device Nodes in /dev.
[  OK  ] Reached target Local File Systems (Pre).
         Starting udev Kernel Device Manager...
[  OK  ] Started Load/Save Random Seed.
[  OK  ] Started udev Kernel Device Manager.
[  OK  ] Started Flush Journal to Persistent Storage.
[  OK  ] Started Rebuild Hardware Database.
         Starting udev Coldplug all Devices...
```

图 1-33　安装过程

第二十步，语言默认选择 English，单击“Continue”按钮，如图 1-34 所示。

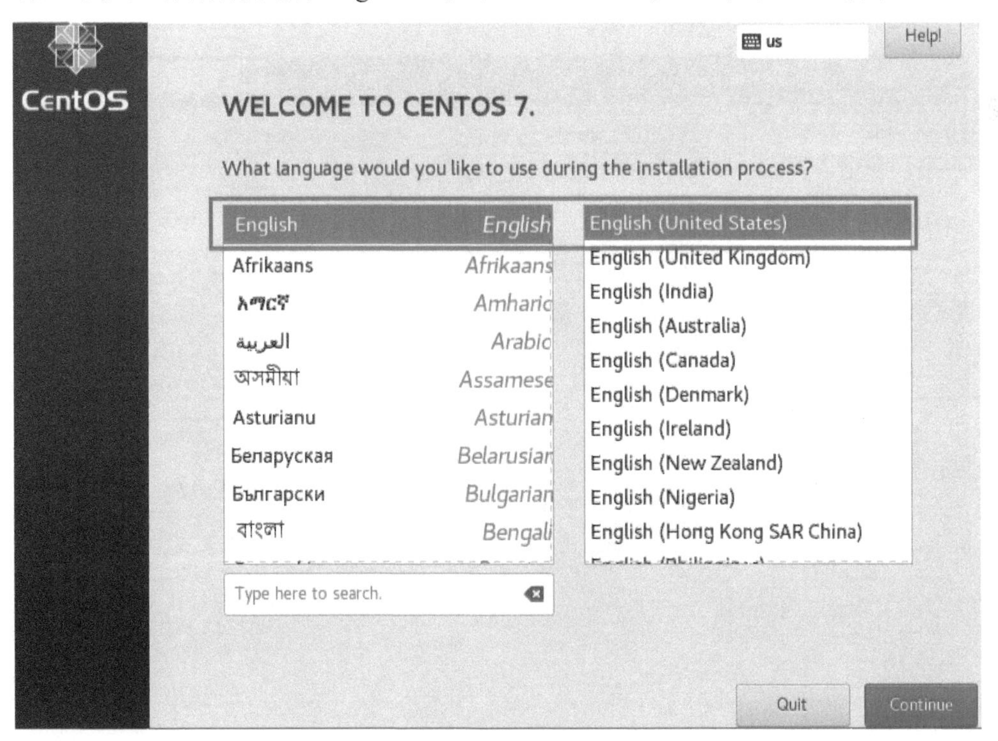

图 1-34　选择语言

第二十一步，在安装信息界面，单击“DATE & TIME”设置时区，如图 1-35 所示。

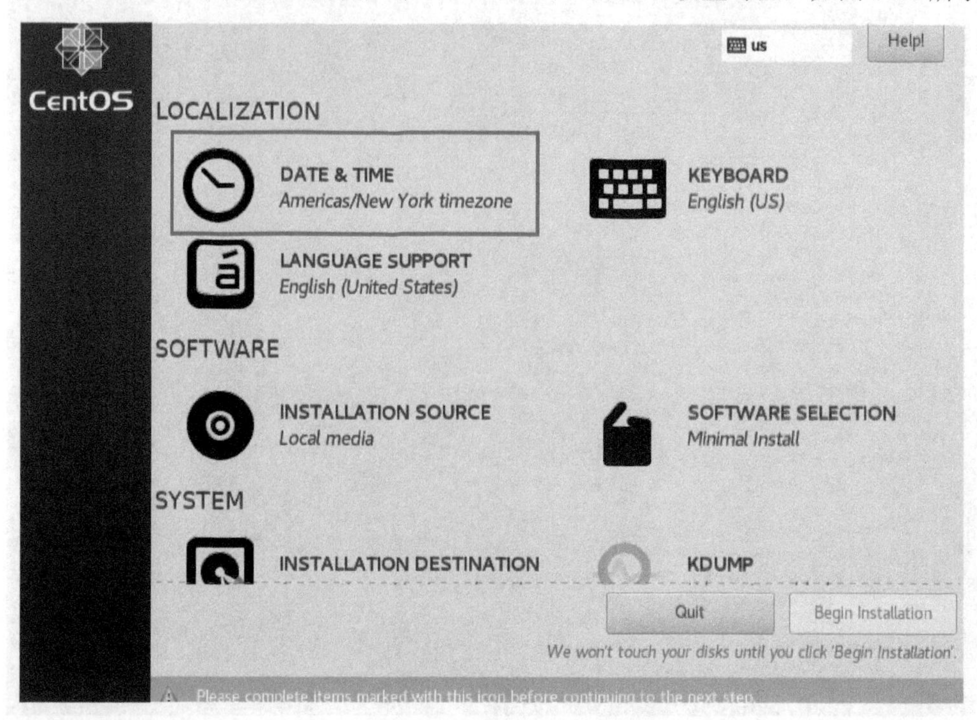

图 1-35　安装信息界面

第二十二步，“Region”(地区)选择“Asia”，“City”(城市)选择“Shanghai”，单击“Done”按钮，如图 1-36 所示。

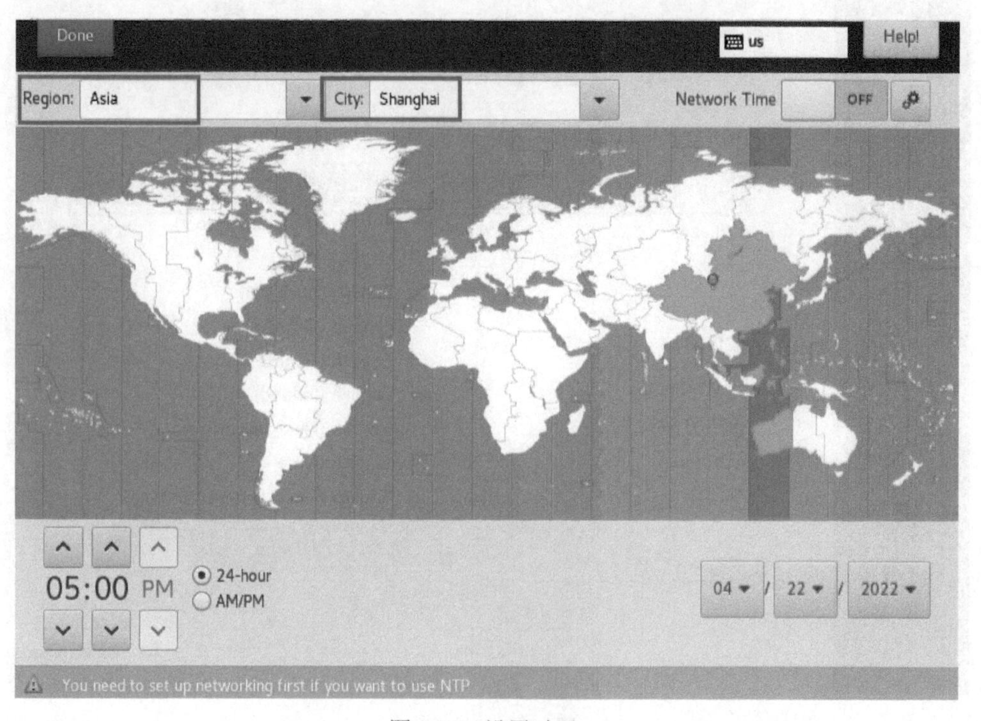

图 1-36　设置时区

第二十三步，在安装信息界面，单击“SOFTWARE SELECTION”选择软件，如图 1-37 所示。

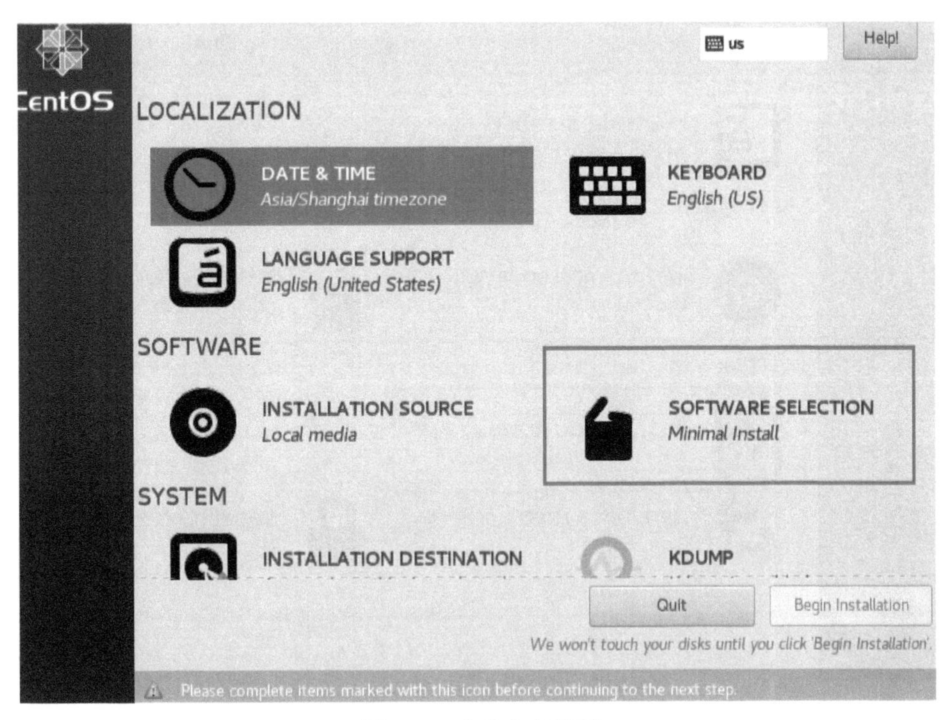

图 1-37　安装信息界面

第二十四步，基础环境选择“Minimal Install”，附加组件全选，单击“Done”按钮，如图 1-38 所示。

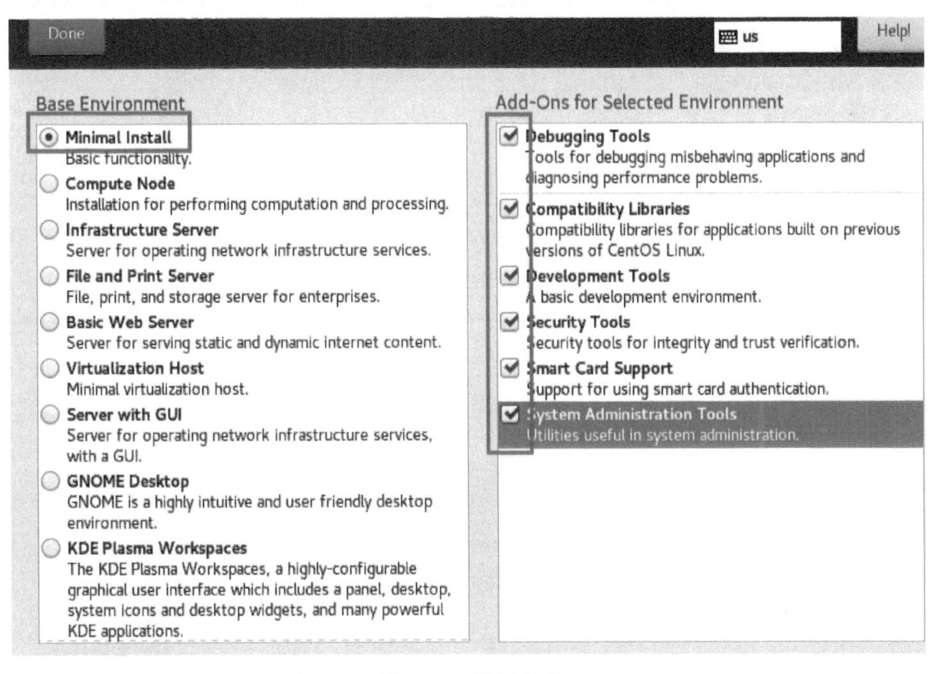

图 1-38　选择软件

第二十五步，在安装信息界面，单击“INSTALLATION DESTINATION”设置系统分区，如图 1-39 所示。

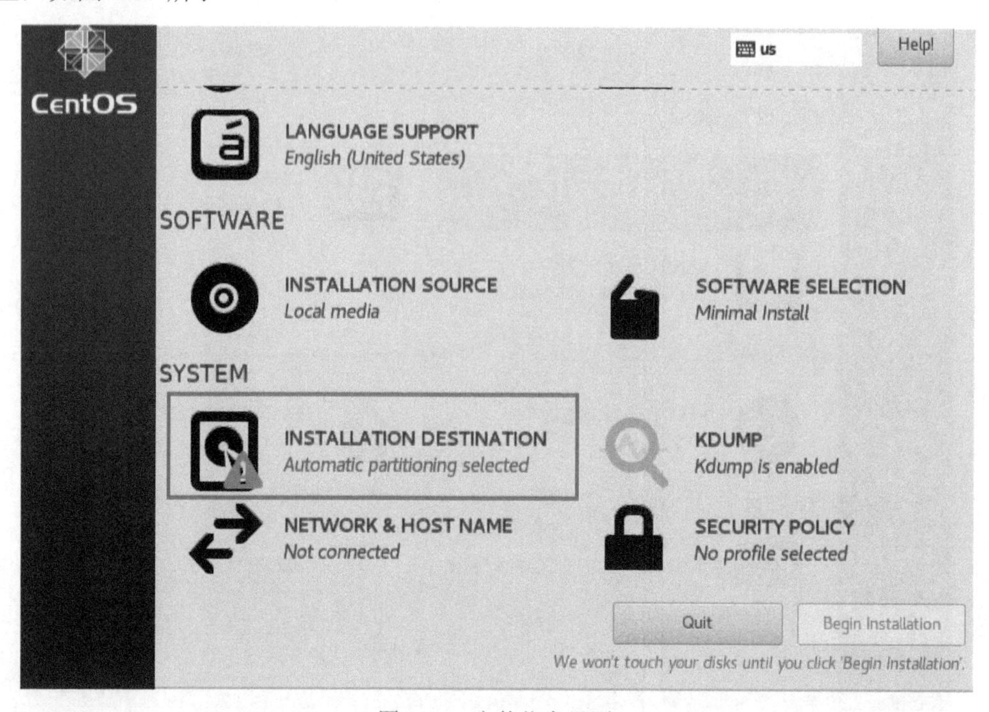

图 1-39　安装信息界面

第二十六步，默认选择自动分区，单击“Done”按钮，如图 1-40 所示。

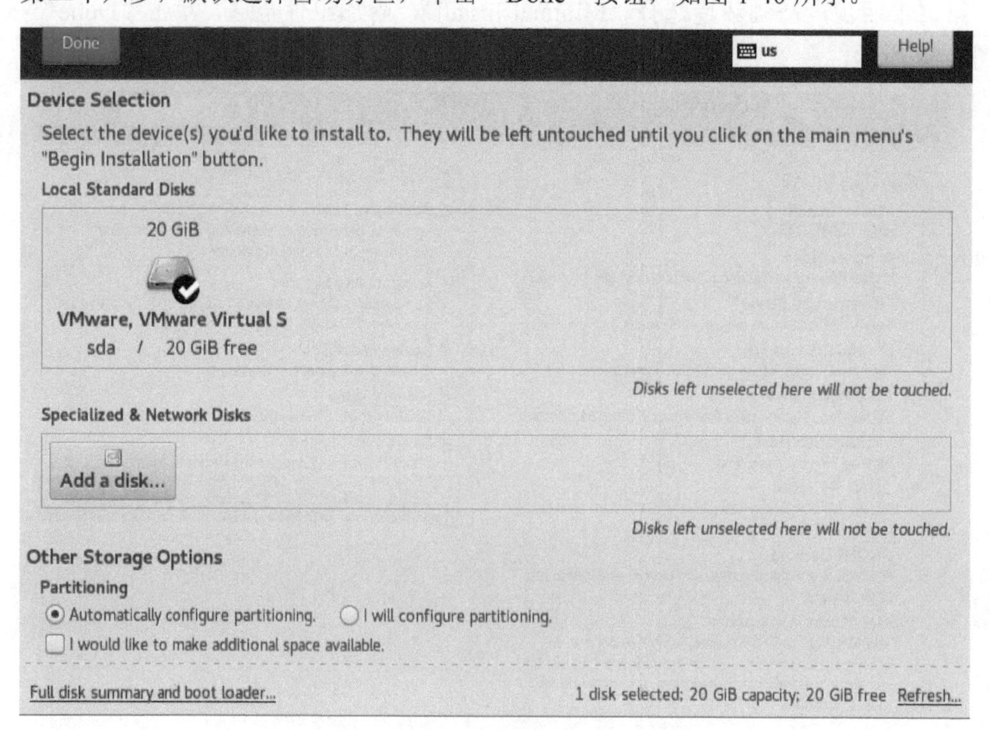

图 1-40　系统分区界面

第二十七步，在安装信息界面，单击“NETWORK & HOST NAME”设置网络，如图 1-41 所示。

图 1-41　安装信息界面

第二十八步，滑动滑块，开启网络服务，注意此时的主机名称默认为“localhost.localdomain”，网卡的名称为“ens33”，单击“Done”按钮，如图 1-42 所示。

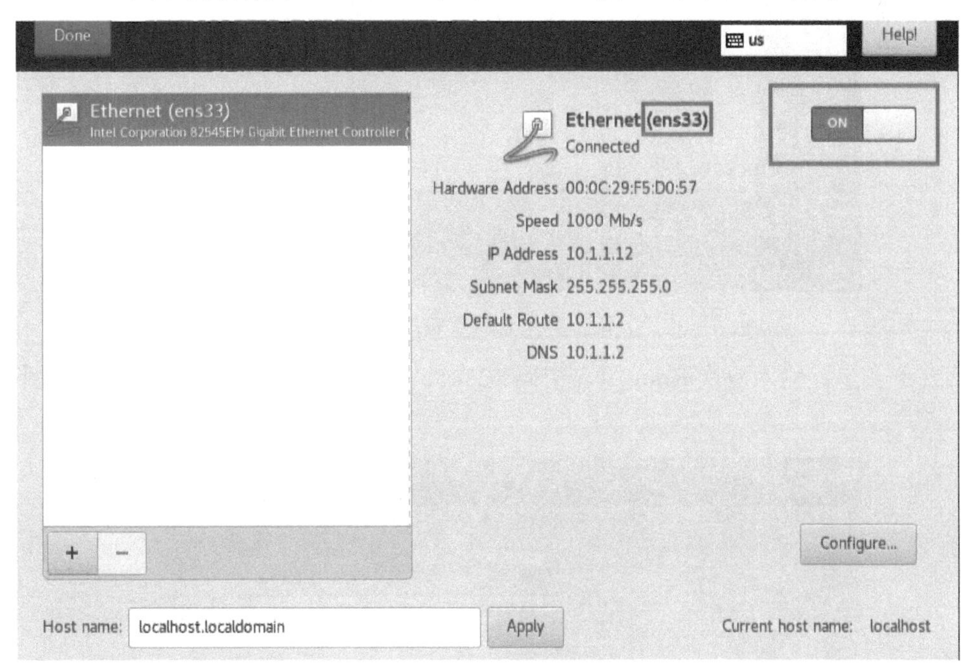

图 1-42　设置网络

第二十九步，在安装信息界面，单击“Begin Installation”按钮，如图 1-43 所示。

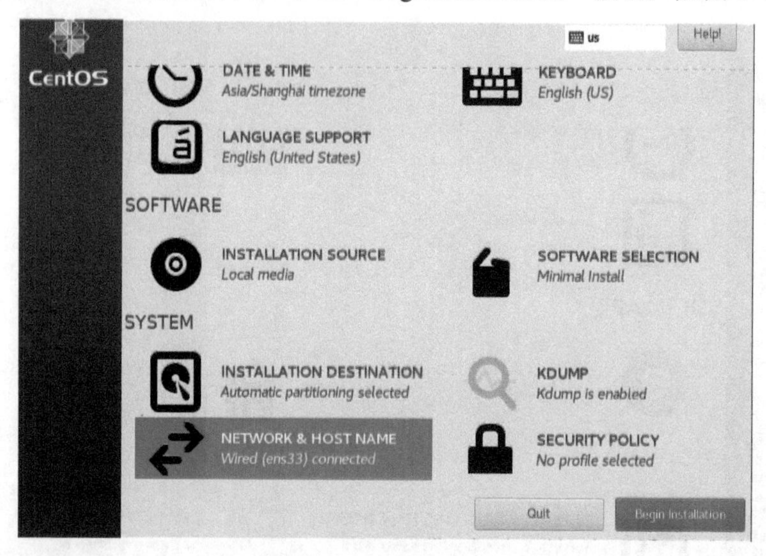

图 1-43　安装信息界面

第三十步，在系统安装界面，单击”ROOT PASSWORD”，设置 Root 用户密码，如图 1-44 所示。

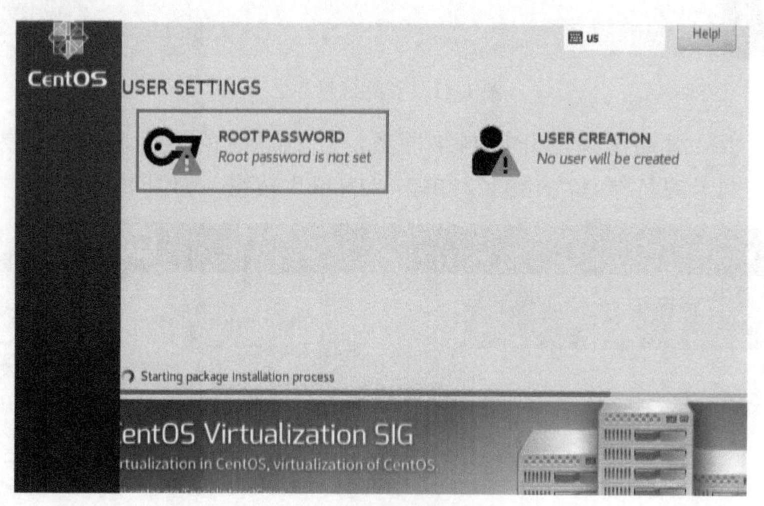

图 1-44　系统安装界面

第三十一步，输入两次 Root 用户密码 123456，单击两次“Done”按钮，如图 1-45 所示。

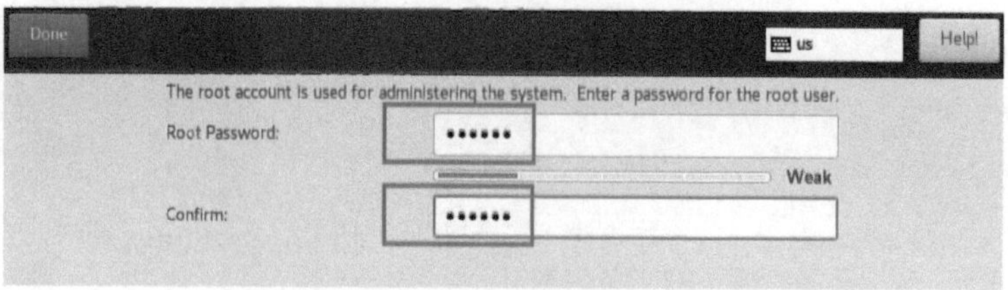

图 1-45　输入 Root 用户密码

第三十二步，等待 CentOS 系统安装，如图 1-46 所示。

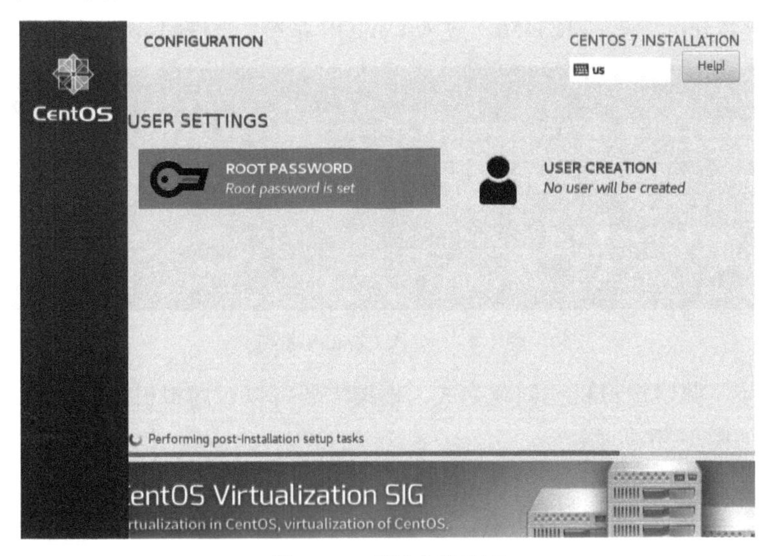

图 1-46　系统安装界面

第三十三步，CentOS 系统安装完成，单击“Reboot”按钮重启系统，如图 1-47 所示。

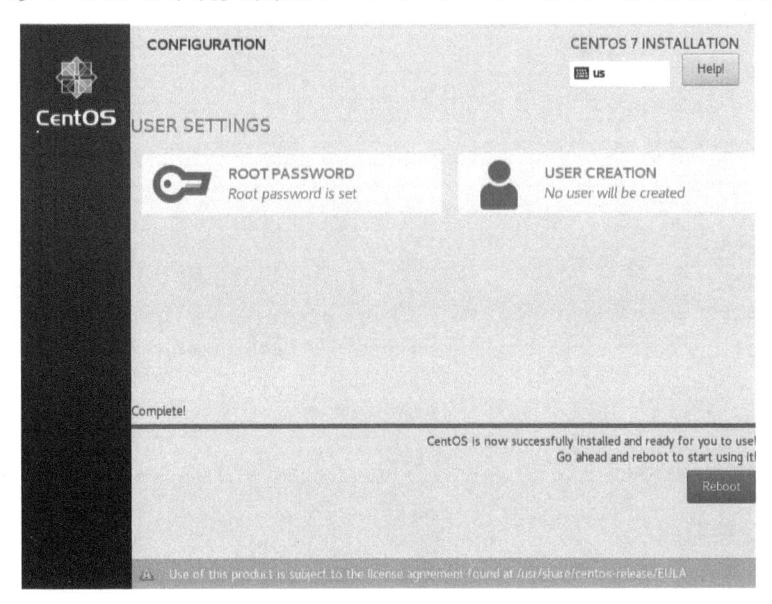

图 1-47　完成系统安装

第三十四步，CentOS 系统重启完成，进入系统操作界面，如图 1-48 所示。

```
CentOS Linux 7 (Core)
Kernel 3.10.0-957.el7.x86_64 on an x86_64

localhost login:
```

图 1-48　系统操作界面

第三十五步，用鼠标在黑色背景区域单击，进入操作系统界面(此时看不到鼠标图标)，输入用户名 root，密码 123456，进入 CentOS 系统，如图 1-49 所示。

```
CentOS Linux 7 (Core)
Kernel 3.10.0-957.el7.x86_64 on an x86_64

localhost login: root
Password:
[root@localhost ~]# _
```

图 1-49　进入 CentOS 系统

第三十六步，此时虚拟机已创建完成，根据需要可以对虚拟机拍摄快照，保存虚拟机此时的状态，如图 1-50 所示。

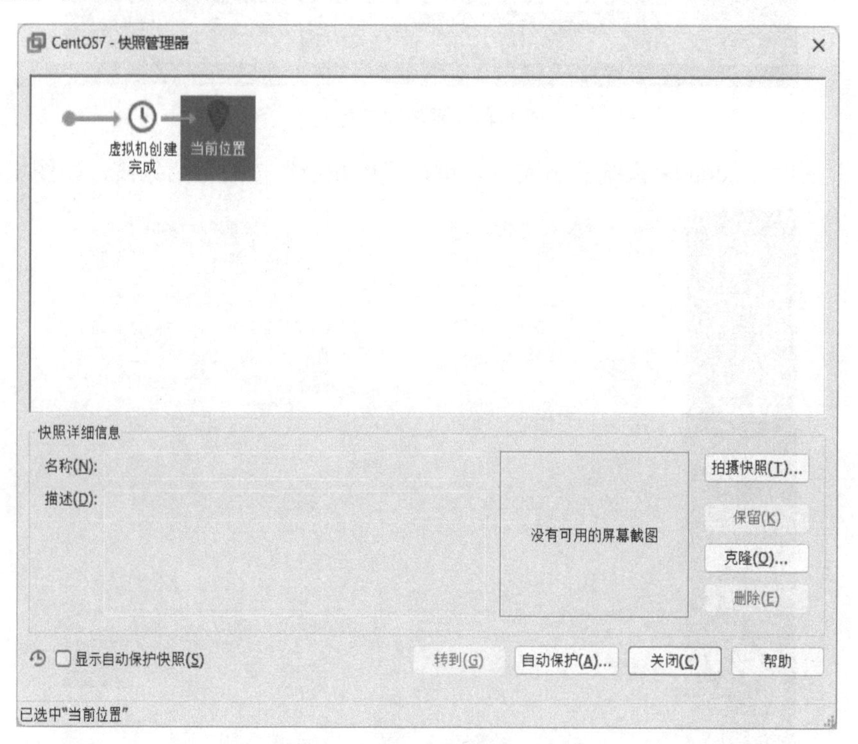

图 1-50　拍摄快照“虚拟机创建完成”

1.2.3　配置虚拟机网络

配置虚拟机网络

1. 任务目标

掌握配置虚拟机网络的方法。

2. 任务内容

(1) 更改主机名。

(2) 配置虚拟机网络。

3. 完成任务所需的设备和软件

(1) 一台安装 Windows 10 操作系统的计算机。

(2) VMware Workstation。

(3) CentOS7.6 镜像。

4. 服务器规划

服务器规划如表 1-2 所示。

表 1-2　服务器规划

编号	主机名称	IP 地址	角色
1	docker	192.168.1.10	Docker 服务器

5. 任务实施步骤

第一步，更改主机名称为 docker，操作命令如下：

```
[root@localhost ~]# hostname
[root@localhost ~]# hostnamectl set-hostname docker
[root@localhost ~]# su
[root@docker ~]# hostname
```

命令运行结果如图 1-51 所示。

```
[root@localhost ~]# hostname
localhost.localdomain
[root@localhost ~]# hostnamectl set-hostname docker
[root@localhost ~]# su
[root@docker ~]# hostname
docker
[root@docker ~]#
```

图 1-51　更改主机名称

第二步，配置静态 IP 地址，操作命令如下：

```
[root@docker ~]# vi /etc/sysconfig/network-scripts/ifcfg-ens33
TYPE="Ethernet"
BOOTPROTO="static"
IPADDR=192.168.1.10
NETMASK=255.255.255.0
GATEWAY=192.168.1.2
NAME="ens33"
DEVICE="ens33"
ONBOOT="yes"
DNS1=114.114.114.114
DNS2=8.8.8.8
```

第三步，选择“我的计算机”下的 CentOS7，单击菜单中的命令“挂起客户机”，如图 1-52 所示。

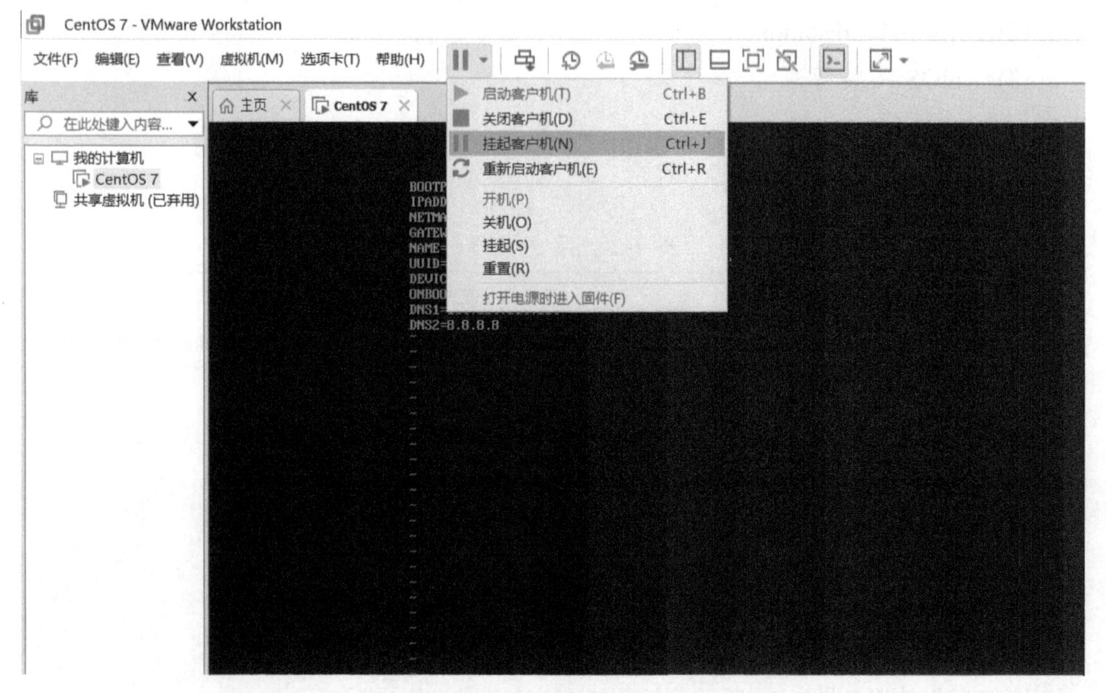

图 1-52　挂起客户机

第四步，单击“编辑”菜单中的“虚拟网络编辑器”，如图 1-53 所示。

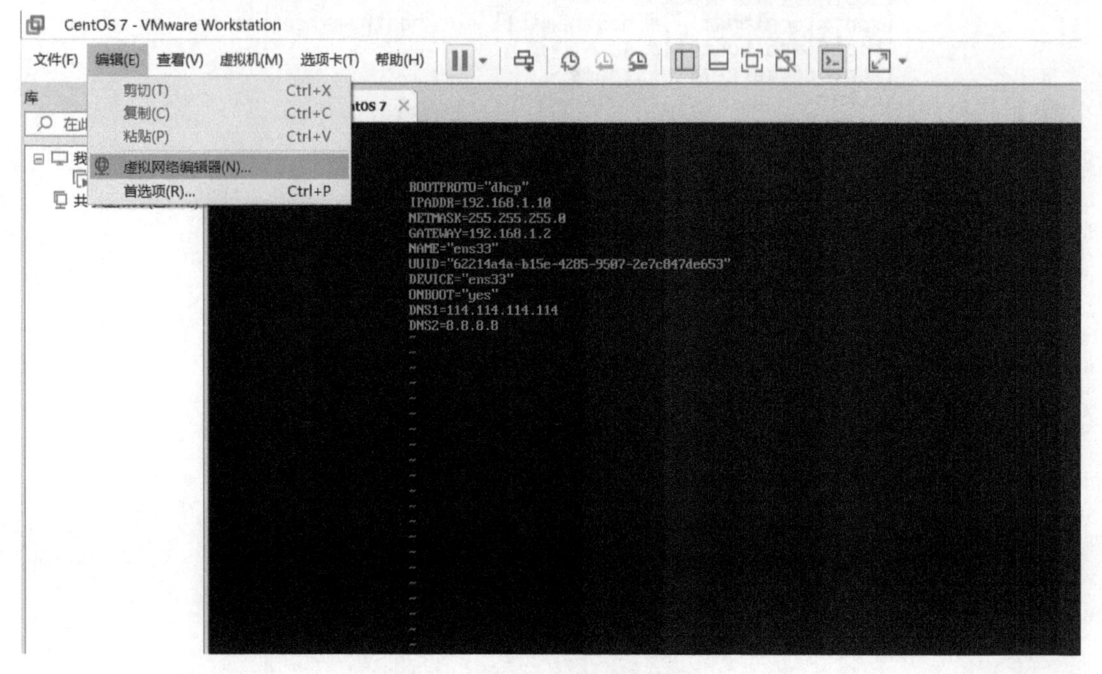

图 1-53　打开“虚拟网络编辑器”

第五步，在“虚拟网络编辑器”对话框中，单击“更改设置”按钮，如图 1-54 所示。

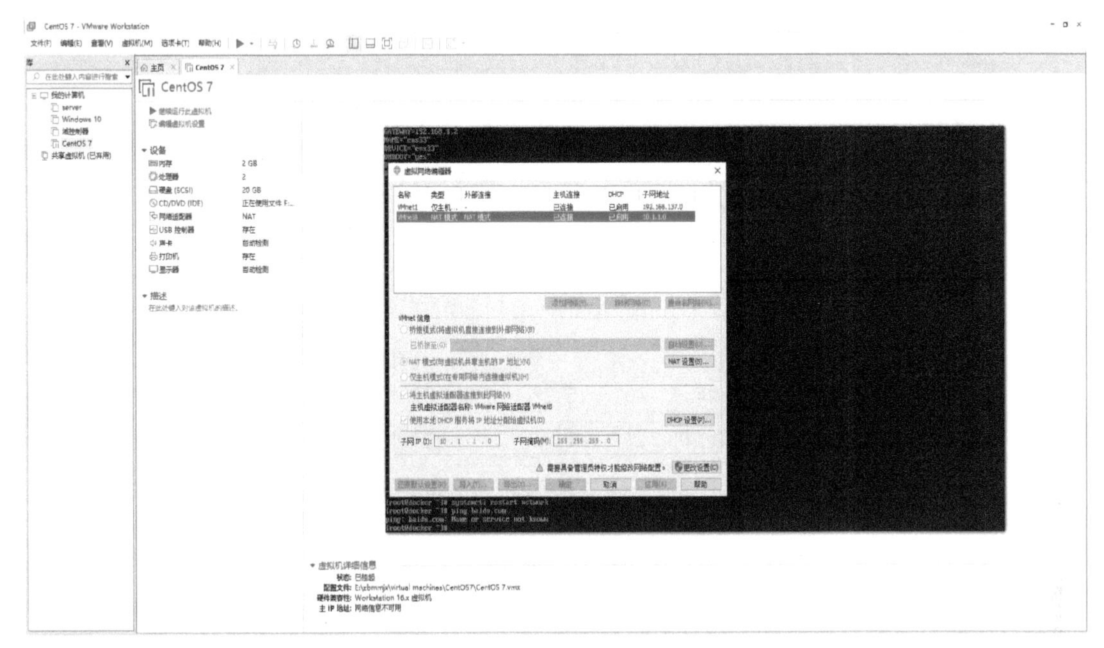

图 1-54　“虚拟网络编辑器”对话框

第六步，在“虚拟网络编辑器”对话框中，选择“VMnet8”，将子网 IP 修改为 192.168.1.0，子网掩码为 255.255.255.0，如图 1-55 所示。

图 1-55　修改 VMnet8 的子网 IP

第七步，在“虚拟网络编辑器”对话框中单击“DHCP 设置”按钮，将起始 IP 地址修改为 192.168.1.128，将结束 IP 地址修改为 192.168.1.254，然后单击“确定”按钮，如

图1-56所示。

DHCP设置

网络: vmnet8

子网IP: 192.168.1.0

子网掩码: 255.255.255.0

起始IP地址(S): 192.168.1.128

结束IP地址(E): 192.168.1.254

广播地址: 192.168.1.255

	天:	小时:	分钟:
默认租用时间(D):	0	0	30
最长租用时间(M):	0	2	0

确定 取消 帮助

图1-56 设置DHCP

第八步，在“虚拟网络编辑器”对话框中单击“NAT设置”按钮，将网关IP修改为192.168.1.2，然后单击“确定”按钮，如图1-57所示。

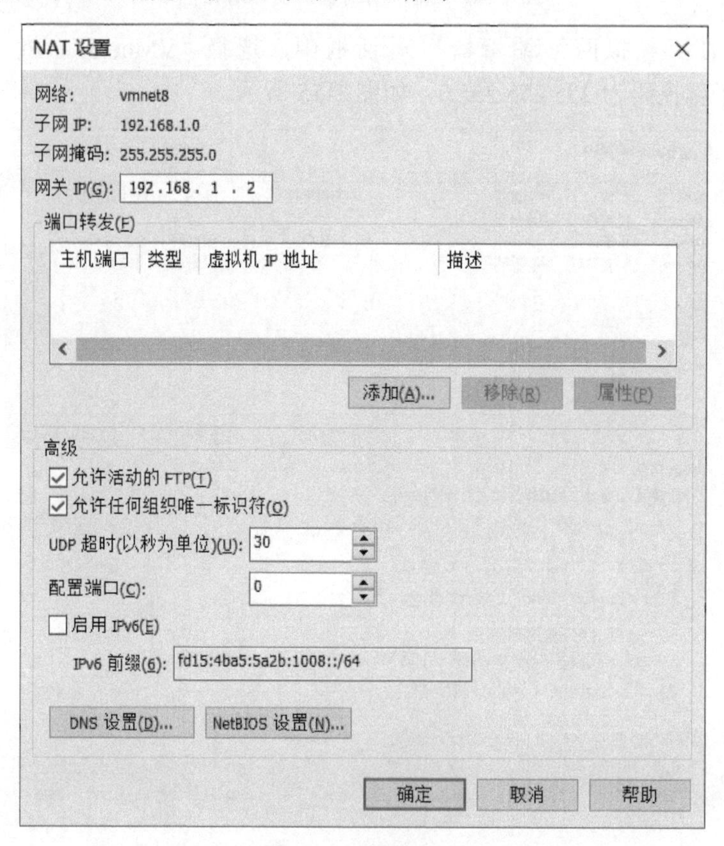

图1-57 设置NAT

第九步，在“虚拟网络编辑器”对话框中单击“应用”按钮，最后单击“确定”按钮，如图1-58所示。

图 1-58 虚拟网络编辑完成

第十步，在“VMware Workstation”窗口单击“继续运行此虚拟机”，如图 1-59 所示。

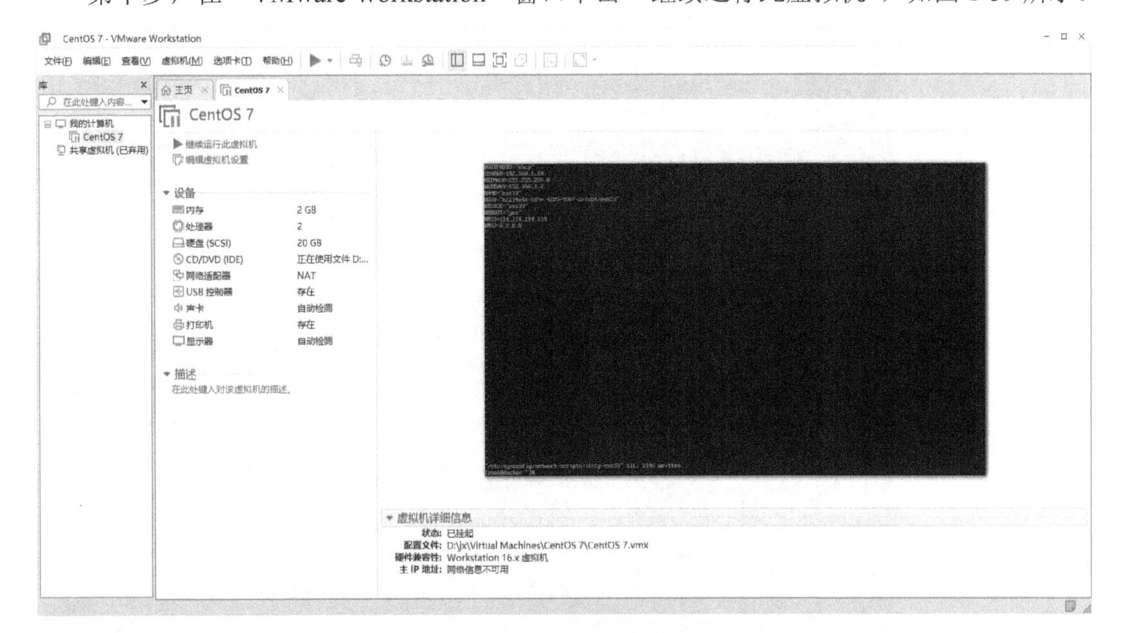

图 1-59 继续运行虚拟机 CentOS7

第十一步，重启虚拟机 CentOS7 的网络服务，并查看 IP 地址，操作命令如下：

```
[root@docker ~]# systemctl restart network
[root@docker ~]# ip a
```

命令运行结果如图 1-60 所示，可以看出网卡 ens33 的 IP 地址为 192.168.1.10。

```
[root@docker ~]# systemctl restart network
[root@docker ~]# ip a
1: lo: <LOOPBACK,UP,LOWER_UP> mtu 65536 qdisc noqueue state UNKNOWN group default qlen 1000
    link/loopback 00:00:00:00:00:00 brd 00:00:00:00:00:00
    inet 127.0.0.1/8 scope host lo
       valid_lft forever preferred_lft forever
    inet6 ::1/128 scope host
       valid_lft forever preferred_lft forever
2: ens33: <BROADCAST,MULTICAST,UP,LOWER_UP> mtu 1500 qdisc pfifo_fast state UP group default qlen 10
00
    link/ether 00:0c:29:43:eb:50 brd ff:ff:ff:ff:ff:ff
    inet 192.168.1.10/24 brd 192.168.1.255 scope global noprefixroute ens33
       valid_lft forever preferred_lft forever
    inet6 fe80::20c:29ff:fe43:eb50/64 scope link
       valid_lft forever preferred_lft forever
[root@docker ~]#
```

图 1-60 重启网卡，并查看 IP 地址

第十二步，测试虚拟机与外网的连通性，操作命令如下：

```
[root@docker ~]# ping www.baidu.com
```

命令运行结果如图 1-61 所示。

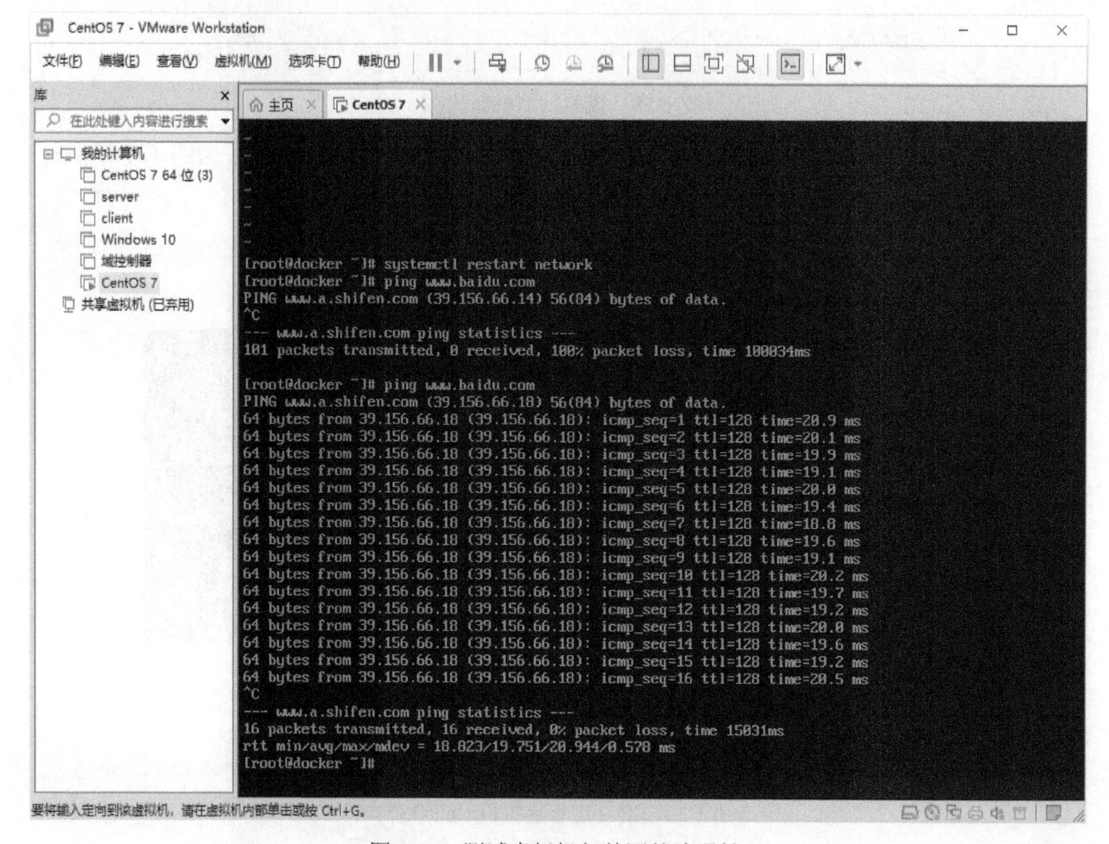

图 1-61 测试虚拟机与外网的连通性

可见，虚拟机与外网是连通的。

第十三步，在 Windows 的命令提示符窗口，测试物理机与此虚拟机的网络是否连通，

如图 1-62 所示，网络已连通。

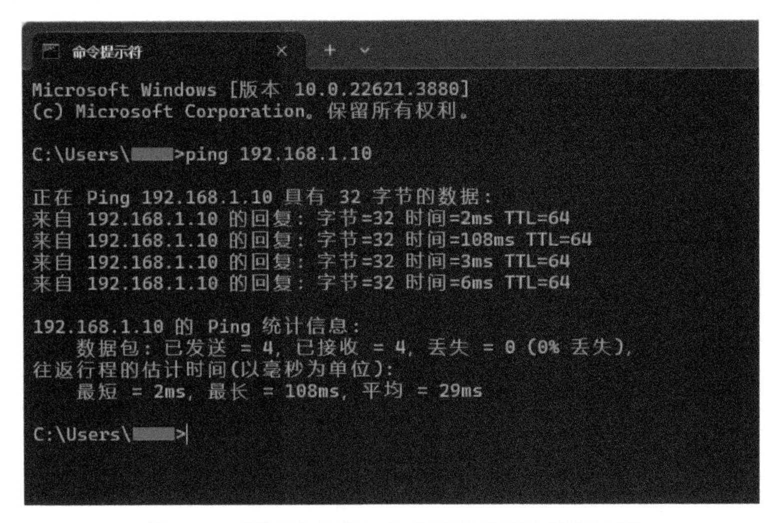

图 1-62　测试物理机与此虚拟机的网络是否连通

第十四步，此时虚拟机 CentOS7 的网络配置完毕，根据需要可以对虚拟机拍摄快照，保存虚拟机此时的状态，如图 1-63 所示。

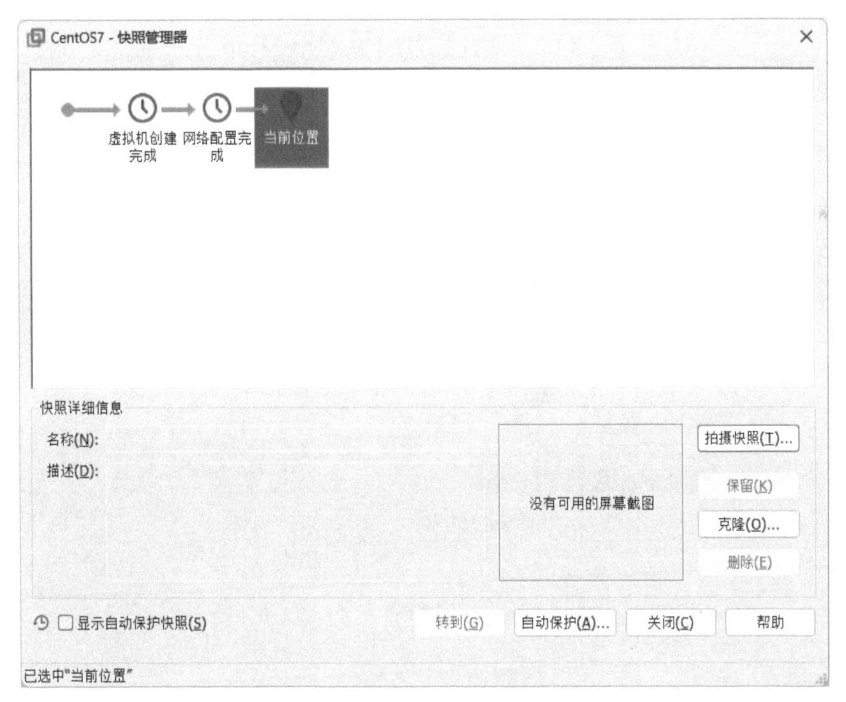

图 1-63　创建快照“网络配置完成”

1.2.4　虚拟机连接远程管理工具

虚拟机连接远程管理工具

1. 任务目标

掌握通过远程管理工具管理虚拟机的方法。

2. 任务内容

(1) 使用虚拟机连接远程管理工具。

(2) 使用远程管理工具 MobaXterm。

3. 完成任务所需的设备和软件

(1) 一台安装 Windows 10 操作系统的计算机。

(2) VMware Workstation。

(3) 远程管理工具 MobaXterm 汉化包。

4. 任务实施步骤

第一步，从附带资源中下载远程管理工具 MobaXterm 的汉化包并将其解压，打开 MobaXterm_CHS.exe，如图 1-64 所示。

Custom.mxtpro	2020/2/2 15:07	MXTPRO 文件	4 KB
CygUtils.plugin	2020/2/2 15:27	PLUGIN 文件	18,825 KB
MobaXterm backup.zip	2024/5/11 11:47	360压缩 ZIP 文件	46 KB
MobaXterm.ini	2024/8/2 10:22	配置设置	8 KB
MobaXterm_CHS.exe	2020/2/2 14:48	应用程序	13,929 KB

图 1-64　下载远程管理工具 MobaXterm

第二步，在远程管理工具 MobaXterm 窗口中，单击“会话”按钮，如图 1-65 所示。

图 1-65　远程管理工具 MobaXterm 窗口

第三步，在“会话设置”窗口中，单击“SSH”按钮，在“远程主机”中输入虚拟机的 IP 地址“192.168.1.10”，勾选“指定用户名”，输入“root”，单击“书签设置”，将“会话名称”设置为“Docker”，单击“好的”按钮，如图 1-66 所示。

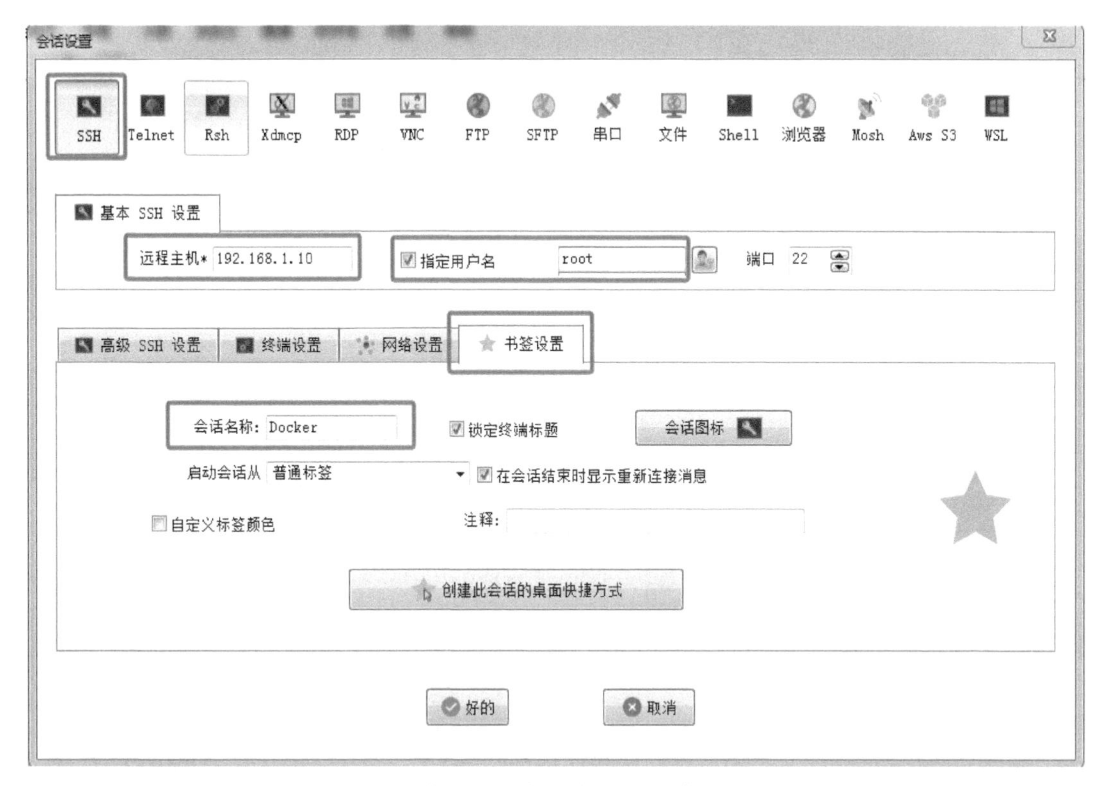

图 1-66　“会话设置”对话框

第四步，在“Docker”会话窗口，输入 root 用户的密码并回车，在弹出的对话框中单击“不”按钮，如图 1-67 所示。

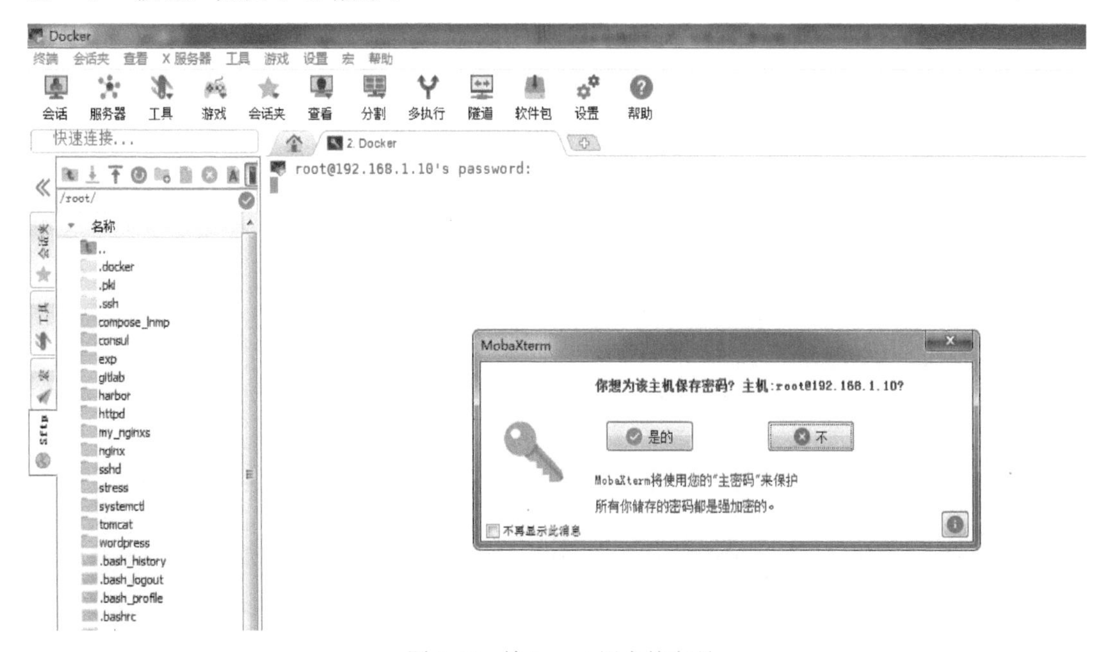

图 1-67　输入 root 用户的密码

第五步，虚拟机连接成功后，即可对虚拟机进行远程管理操作，如图 1-68 所示。

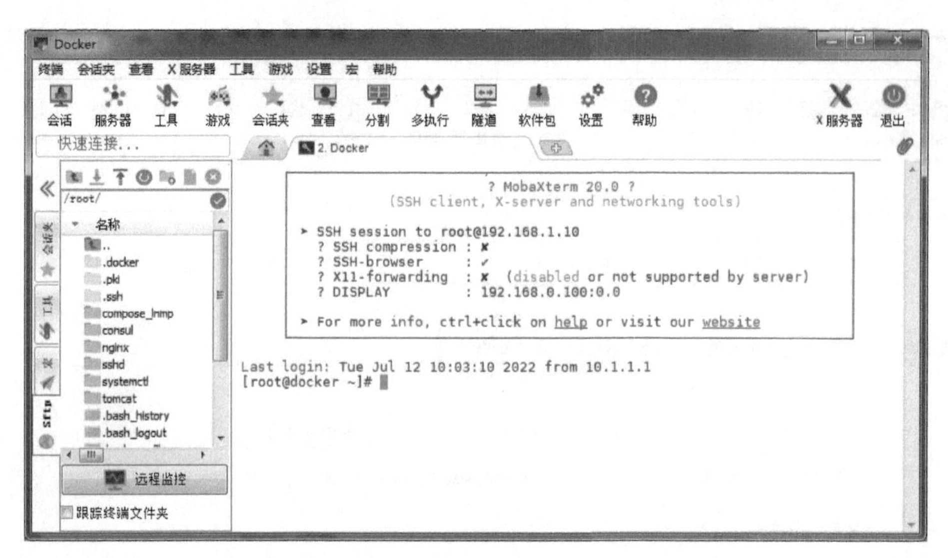

图 1-68 远程管理工具 MobaXterm 成功连接虚拟机

1.2.5 初始化虚拟机

初始化虚拟机

1. 任务目标

掌握初始化虚拟机，创建 Docker 运行环境的方法。

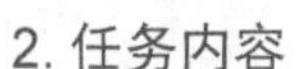

2. 任务内容

(1) 关闭防火墙、SELinux 以及 NetworkManager。

(2) 配置 yum 源。

(3) 安装和设置 ntp 时间同步。

3. 完成任务所需的设备和软件

(1) 一台安装 Windows 10 操作系统的计算机。

(2) VMware Workstation。

(3) 远程管理工具 MobaXterm。

4. 任务实施步骤

第一步，关闭防火墙，操作命令如下：

```
[root@docker ~]# systemctl stop firewalld
[root@docker ~]# systemctl disable firewalld
```

命令运行结果如图 1-69 所示。

```
[root@docker ~]# systemctl stop firewalld
[root@docker ~]# systemctl disable firewalld
Removed symlink /etc/systemd/system/multi-user.target.wants/firewalld.service.
Removed symlink /etc/systemd/system/dbus-org.fedoraproject.FirewallD1.service.
[root@docker ~]#
```

图 1-69 关闭防火墙

第二步，关闭 SELinux，操作命令如下：

```
[root@docker ~]# setenforce 0( 临时关闭 SELinux)
```

```
[root@docker ~]# vi /etc/selinux/config
# This file controls the state of SELinux on the system.
# SELinux= can take one of these three values:
#     enforcing - SELinux security policy is enforced.
#     permissive - SELinux prints warnings instead of enforcing.
#     disabled - No SELinux policy is loaded.
SELinux=disabled
# SELinuxTYPE= can take one of three values:
#     targeted - Targeted processes are protected,
#     minimum - Modification of targeted policy. Only selected processes are protected.
#     mls - Multi Level Security protection.
SELinuxTYPE=targeted
```

注意：SELinux 取值不同则工作状态不同。

SELinux=enforcing，表示执行 SELinux 安全策略，违反策略的行为会被禁止并记录到日志中。

SELinux=permissive，表示执行 SELinux 安全策略，违反策略的行为会被警告而不是禁止。

SELinux=disabled，表示 SELinux 关闭，不强制执行任何策略，仅记录违反策略的行为。

命令运行结果如图 1-70 所示。

```
[root@docker ~]# setenforce 0
[root@docker ~]# vi /etc/selinux/config
[root@docker ~]#
```

图 1-70　关闭 SELinux

第三步，关闭 NetworkManager，操作命令如下：

```
[root@docker ~]# systemctl stop NetworkManager
[root@docker ~]# systemctl disable NetworkManager
```

命令运行结果如图 1-71 所示。

```
[root@docker ~]# systemctl stop NetworkManager
[root@docker ~]# systemctl disable NetworkManager
Removed symlink /etc/systemd/system/multi-user.target.wants/NetworkManager.service.
Removed symlink /etc/systemd/system/dbus-org.freedesktop.NetworkManager.service.
Removed symlink /etc/systemd/system/dbus-org.freedesktop.nm-dispatcher.service.
Removed symlink /etc/systemd/system/network-online.target.wants/NetworkManager-wait-online.service.
[root@docker ~]#
```

图 1-71　关闭 NetworkManager

第四步，配置 yum 源。由于虚拟机默认 yum 源使用官方仓库，导致访问速度很慢甚至不可用，因此需更将其换为国内的 yum 源。此处以阿里云为例介绍如何配置 yum 源。

(1) 备份系统旧配置文件，操作命令如下：

```
[root@docker ~]# mv /etc/yum.repos.d/CentOS-Base.repo /etc/yum.repos.d/CentOS-Base.repo.backup
[root@docker ~]# ls /etc/yum.repos.d/
```

命令运行结果如图 1-72 所示。

```
[root@docker ~]# ls /etc/yum.repos.d/
CentOS-Base.repo  CentOS-Debuginfo.repo  CentOS-Media.repo    CentOS-Vault.repo
CentOS-CR.repo    CentOS-fasttrack.repo  CentOS-Sources.repo
[root@docker ~]# mv /etc/yum.repos.d/CentOS-Base.repo /etc/yum.repos.d/CentOS-Base.repo.backup
[root@docker ~]# ls /etc/yum.repos.d/
CentOS-Base.repo.backup  CentOS-Debuginfo.repo  CentOS-Media.repo    CentOS-Vault.repo
CentOS-CR.repo           CentOS-fasttrack.repo  CentOS-Sources.repo
[root@docker ~]#
```

图 1-72　备份系统旧配置文件

(2) 通过网址 https://mirrors.aliyun.com/repo/Centos-7.repo 下载文件 Centos-7.repo(或从随书附带资源中下载)，将其重命名为 CentOS-Base.repo 之后拖拽至虚拟机的 /etc/yum.repos.d/ 目录下，如图 1-73 所示。

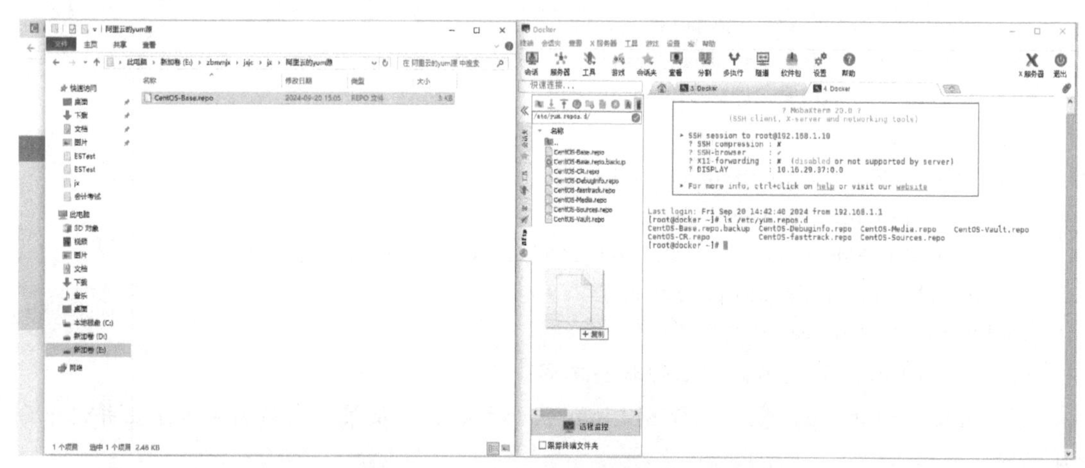

图 1-73　将 CentOS-Base.repo 拖拽至虚拟机的 /etc/yum.repos.d/ 目录下

(3) 更新缓存，操作命令如下：

```
[root@docker ~]# yum clean all
[root@docker ~]# yum makecache
```

命令运行结果如图 1-74 所示。

```
[root@docker ~]# yum clean all
Loaded plugins: fastestmirror
Cleaning repos: base extras updates
[root@docker ~]# yum makecache
Loaded plugins: fastestmirror
Determining fastest mirrors
 * base: mirrors.aliyun.com
 * extras: mirrors.aliyun.com
 * updates: mirrors.aliyun.com
base                                                                 | 3.6 kB  00:00:00
extras                                                               | 2.9 kB  00:00:00
updates                                                              | 2.9 kB  00:00:00
(1/10): base/7/x86_64/group_gz                                       | 153 kB  00:00:00
(2/10): extras/7/x86_64/filelists_db                                 | 305 kB  00:00:00
(3/10): base/7/x86_64/other_db                                       | 2.6 MB  00:00:00
(4/10): extras/7/x86_64/other_db                                     | 154 kB  00:00:00
(5/10): updates/7/x86_64/filelists_db                                |  15 MB  00:00:00
(6/10): updates/7/x86_64/other_db                                    | 1.6 MB  00:00:00
(7/10): updates/7/x86_64/primary_db                                  |  27 MB  00:00:00
base/7/x86_64/filelists_db      FAILED
http://mirrors.aliyuncs.com/centos/7/os/x86_64/repodata/d6d94c7d406fe7ad4902a97104b39a0d8299451832a97f31d71653ba982c955b-fil
elists.sqlite.bz2: [Errno 14] curl#7 - "Failed connect to mirrors.aliyuncs.com:80; Connection refused"
Trying other mirror.
base/7/x86_64/primary_db        FAILED
http://mirrors.cloud.aliyuncs.com/centos/7/os/x86_64/repodata/6d0c3a488c282fe537794b5946b01e28c7f44db79097bb06826e1c0c88bad5
ef-primary.sqlite.bz2: [Errno 14] curl#7 - "Failed connect to mirrors.cloud.aliyuncs.com:80; Connection refused"
Trying other mirror.
extras/7/x86_64/primary_db      FAILED
http://mirrors.aliyuncs.com/centos/7/extras/x86_64/repodata/e12dbf10e94bc2b33b1f45e026559bc8685728b139dddae0654d96bc624c5602
-primary.sqlite.bz2: [Errno 14] curl#7 - "Failed connect to mirrors.aliyuncs.com:80; Connection refused"
Trying other mirror.
(8/10): base/7/x86_64/filelists_db                                             | 7.2 MB  00:00:00
(9/10): extras/7/x86_64/primary_db                                             | 253 kB  00:00:00
(10/10): base/7/x86_64/primary_db                                              | 6.1 MB  00:00:00
Metadata Cache Created
[root@docker ~]#
```

图 1-74　更新缓存

注意：对于以上结果中的 FAILED 信息，可通过启动 NetworkManager、重启网络和重新更新缓存得到解决，操作命令如下：

```
[root@docker ~]# systemctl start NetworkManager
[root@docker ~]# systemctl restart network
[root@docker ~]# yum clean all
```

[root@docker ~]# yum makecache

第五步，安装vim编辑器和wget下载工具，操作命令如下：

[root@docker ~]# yum install vim wget -y

命令运行结果如图1-75所示。

```
[root@docker ~]# yum install vim wget -y
Loaded plugins: fastestmirror
Loading mirror speeds from cached hostfile
 * base: mirrors.aliyun.com
 * extras: mirrors.aliyun.com
 * updates: mirrors.aliyun.com
Resolving Dependencies
--> Running transaction check
---> Package vim-enhanced.x86_64 2:7.4.629-8.el7_9 will be installed
--> Processing Dependency: vim-common = 2:7.4.629-8.el7_9 for package: 2:vim-enhanced-7.4.629-8.el7_9.x86_64
--> Processing Dependency: libgpm.so.2()(64bit) for package: 2:vim-enhanced-7.4.629-8.el7_9.x86_64
---> Package wget.x86_64 0:1.14-18.el7_6.1 will be installed
--> Running transaction check
---> Package gpm-libs.x86_64 0:1.20.7-6.el7 will be installed
---> Package vim-common.x86_64 2:7.4.629-8.el7_9 will be installed
--> Processing Dependency: vim-filesystem for package: 2:vim-common-7.4.629-8.el7_9.x86_64
--> Running transaction check
---> Package vim-filesystem.x86_64 2:7.4.629-8.el7_9 will be installed
--> Finished Dependency Resolution

Dependencies Resolved

================================================================================
 Package            Arch        Version                 Repository        Size
================================================================================
Installing:
 vim-enhanced       x86_64      2:7.4.629-8.el7_9       updates          1.1 M
 wget               x86_64      1.14-18.el7_6.1         base             547 k
Installing for dependencies:
 gpm-libs           x86_64      1.20.7-6.el7            base              32 k
 vim-common         x86_64      2:7.4.629-8.el7_9       updates          5.9 M
 vim-filesystem     x86_64      2:7.4.629-8.el7_9       updates           11 k

Transaction Summary
================================================================================
Install  2 Packages (+3 Dependent packages)

Total download size: 7.5 M
Installed size: 25 M
Downloading packages:
warning: /var/cache/yum/x86_64/7/base/packages/gpm-libs-1.20.7-6.el7.x86_64.rpm: Header V3 RSA/SHA256 Signature, key ID f4a80eb5: NOKEY|   0 B  --:--:-- ETA
Public key for gpm-libs-1.20.7-6.el7.x86_64.rpm is not installed
(1/5): gpm-libs-1.20.7-6.el7.x86_64.rpm                                  |  32 kB  00:00:00
  ......
Complete!
[root@docker ~]#
```

图1-75　安装vim编辑器和wget下载工具

第六步，配置扩展EPEL源。

(1) 下载EPEL到本地/etc/yum.repos.d目录，操作命令如下：

[root@docker ~]# wget -O /etc/yum.repos.d/epel.repo http://mirrors.aliyun.com/repo/epel-7.repo

命令运行结果如图1-76所示。

```
[root@docker ~]# wget -O /etc/yum.repos.d/epel.repo http://mirrors.aliyun.com/repo/epel-7.repo
--2024-08-25 06:54:07--  http://mirrors.aliyun.com/repo/epel-7.repo
Resolving mirrors.aliyun.com (mirrors.aliyun.com)... 182.40.42.231, 182.40.42.228, 140.249.109.232, ...
Connecting to mirrors.aliyun.com (mirrors.aliyun.com)|182.40.42.231|:80... connected.
HTTP request sent, awaiting response... 200 OK
Length: 664 [application/octet-stream]
Saving to: '/etc/yum.repos.d/epel.repo'

100%[=====================================================================================>] 664         --.-K/s   in 0s

2024-08-25 06:54:23 (42.4 MB/s) - '/etc/yum.repos.d/epel.repo' saved [664/664]

[root@docker ~]#
```

图1-76　下载EPEL到本地/etc/yum.repos.d目录

(2) 更新缓存，操作命令如下：

[root@docker ~]# yum clean all

[root@docker ~]# yum makecache

命令运行结果如图1-77所示。

```
[root@docker ~]# yum clean all
Loaded plugins: fastestmirror
Repository cr is listed more than once in the configuration
Repository fasttrack is listed more than once in the configuration
Cleaning repos: epel extras os updates
Cleaning up list of fastest mirrors
Other repos take up 111 M of disk space (use --verbose for details)
[root@docker ~]# yum makecache
Loaded plugins: fastestmirror
Repository cr is listed more than once in the configuration
Repository fasttrack is listed more than once in the configuration
Determining fastest mirrors
epel                                        | 4.3 kB  00:00:00
extras                                      | 2.9 kB  00:00:00
os                                          | 3.6 kB  00:00:00
updates                                     | 2.9 kB  00:00:00
(1/16): epel/x86_64/filelists_db            |  15 MB  00:00:20
(2/16): epel/x86_64/updateinfo              | 1.0 MB  00:00:00
(3/16): epel/x86_64/prestodelta             |  592 B  00:00:00
(4/16): epel/x86_64/group                   | 399 kB  00:00:24
(5/16): extras/7/x86_64/filelists_db        | 305 kB  00:00:00
(6/16): extras/7/x86_64/primary_db          | 253 kB  00:00:00
(7/16): os/7/x86_64/group_gz                | 153 kB  00:00:00
(8/16): extras/7/x86_64/other_db            | 154 kB  00:00:00
(9/16): os/7/x86_64/filelists_db            | 7.2 MB  00:00:01
(10/16): epel/x86_64/primary_db             | 8.7 MB  00:00:05
(11/16): os/7/x86_64/other_db               | 2.6 MB  00:00:00
(12/16): epel/x86_64/other_db               | 4.1 MB  00:00:02
(13/16): os/7/x86_64/primary_db             | 6.1 MB  00:00:02
(14/16): updates/7/x86_64/filelists_db      |  15 MB  00:00:02
(15/16): updates/7/x86_64/other_db          | 1.6 MB  00:00:00
(16/16): updates/7/x86_64/primary_db        |  27 MB  00:00:05
Metadata Cache Created
[root@docker ~]#
```

图 1-77　更新缓存

(3) 测试 EPEL 源。

① 安装 sl 软件，操作命令如下：

```
[root@docker ~]# yum install sl -y
```

命令运行结果如图 1-78 所示。

```
[root@docker ~]# yum install sl -y
Loaded plugins: fastestmirror
Repository cr is listed more than once in the configuration
Repository fasttrack is listed more than once in the configuration
Loading mirror speeds from cached hostfile
Resolving Dependencies
--> Running transaction check
---> Package sl.x86_64 0:5.02-1.el7 will be installed
--> Finished Dependency Resolution

Dependencies Resolved

================================================================================
 Package          Arch            Version              Repository          Size
================================================================================
Installing:
 sl               x86_64          5.02-1.el7           epel                14 k

Transaction Summary
================================================================================
Install  1 Package

Total download size: 14 k
Installed size: 17 k
Downloading packages:
sl-5.02-1.el7.x86_64.rpm                                |  14 kB  00:00:13
Running transaction check
Running transaction test
Transaction test succeeded
Running transaction
  Installing : sl-5.02-1.el7.x86_64                                       1/1
  Verifying  : sl-5.02-1.el7.x86_64                                       1/1

Installed:
  sl.x86_64 0:5.02-1.el7

Complete!
```

图 1-78　安装 sl 软件

② 运行 sl 软件，操作命令如下：

```
[root@docker ~]# sl
```

命令运行结果如图 1-79 所示。

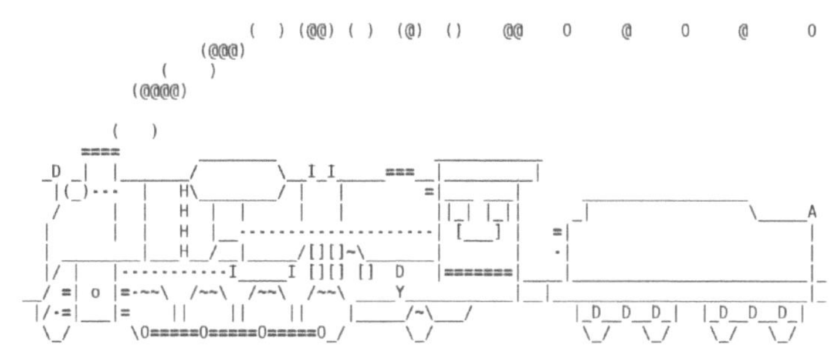

图 1-79　运行 sl 软件

第七步，安装和设置 ntp 时间同步。

① 安装时间同步工具 ntp 和 ntpdate，操作命令如下：

[root@docker ~]# yum install ntp ntpdate -y

命令运行结果如图 1-80 所示。

```
[root@docker ~]# yum install ntp ntpdate -y
Loaded plugins: fastestmirror
Repository cr is listed more than once in the configuration
Repository fasttrack is listed more than once in the configuration
Loading mirror speeds from cached hostfile
Resolving Dependencies
--> Running transaction check
---> Package ntp.x86_64 0:4.2.6p5-29.el7.centos.2 will be installed
--> Processing Dependency: libopts.so.25()(64bit) for package: ntp-4.2.6p5-29.el7.centos.2.x86_64
---> Package ntpdate.x86_64 0:4.2.6p5-29.el7.centos.2 will be installed
--> Running transaction check
---> Package autogen-libopts.x86_64 0:5.18-5.el7 will be installed
--> Finished Dependency Resolution

Dependencies Resolved

================================================================================================================
 Package                    Arch                    Version                         Repository             Size
================================================================================================================
Installing:
 ntp                        x86_64                  4.2.6p5-29.el7.centos.2         os                    549 k
 ntpdate                    x86_64                  4.2.6p5-29.el7.centos.2         os                     87 k
Installing for dependencies:
 autogen-libopts            x86_64                  5.18-5.el7                      os                     66 k

Transaction Summary
================================================================================================================
Install  2 Packages (+1 Dependent package)

Total download size: 701 k
Installed size: 1.6 M
Downloading packages:
(1/3): ntp-4.2.6p5-29.el7.centos.2.x86_64.rpm                                   | 549 kB  00:00:06
(2/3): autogen-libopts-5.18-5.el7.x86_64.rpm                                    |  66 kB  00:00:06
(3/3): ntpdate-4.2.6p5-29.el7.centos.2.x86_64.rpm                               |  87 kB  00:00:00
----------------------------------------------------------------------------------------------------------------
Total                                                                   94 kB/s | 701 kB  00:00:07
Running transaction check
Running transaction test
Transaction test succeeded
Running transaction
  Installing : autogen-libopts-5.18-5.el7.x86_64                                                            1/3
  Installing : ntpdate-4.2.6p5-29.el7.centos.2.x86_64                                                       2/3
……
Complete!
```

图 1-80　安装时间同步工具 ntp 和 ntpdate

② 与时间服务器同步，并设置开机自启时间同步，操作命令如下：

[root@docker ~]# ntpdate 182.92.12.11

[root@docker ~]# systemctl start ntpd

[root@docker ~]# systemctl enable ntpd

命令运行结果如图 1-81 所示。

```
[root@docker ~]#
[root@docker ~]# ntpdate 182.92.12.11
25 Aug 07:47:14 ntpdate[7818]: adjust time server 182.92.12.11 offset 0.131649 sec
[root@docker ~]# systemctl start ntpd
[root@docker ~]# systemctl enable ntpd
Created symlink from /etc/systemd/system/multi-user.target.wants/ntpd.service to /usr/lib/systemd/system/ntpd.service.
[root@docker ~]#
```

图 1-81　与时间服务器同步并开启时间同步

第八步，为虚拟机拍摄快照，然后保存虚拟机此时的状态，如图1-82所示。

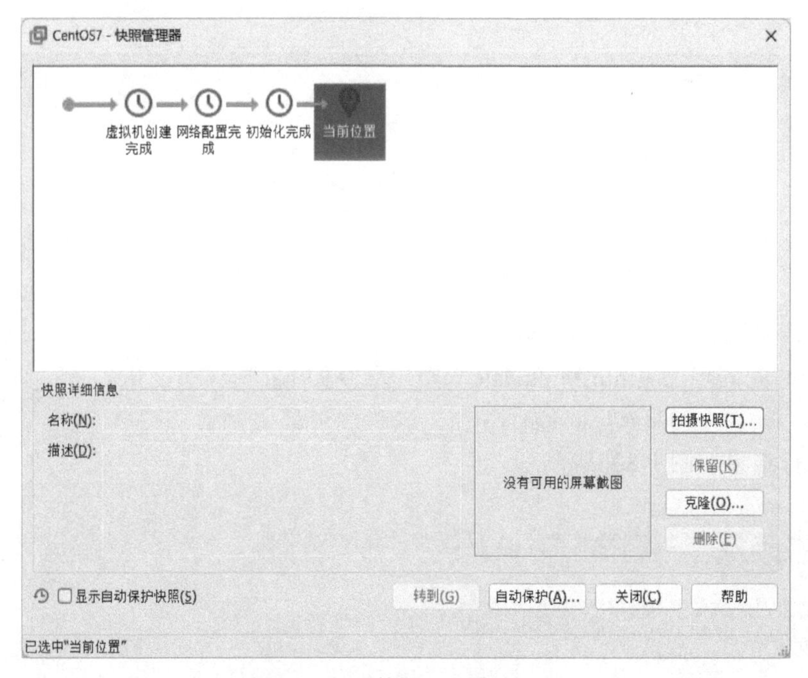

图1-82　拍摄快照“初始化完成”

双创视角

华为鸿蒙系统

党的二十大报告指出，当前世界百年未有之大变局加速演进，新一轮科技革命和产业变革深入发展，国际力量对比深刻调整，我国发展面临新的战略机遇。

2019年8月9日，华为开发者大会(HDC.2019)在东莞举行，会上华为鸿蒙系统(HUAWEI HarmonyOS)正式发布。HarmonyOS是面向万物互联的全场景分布式操作系统，支持多种终端设备运行，如手机、平板、智能穿戴、智慧屏等，是提供应用开发和设备开发的一站式服务平台。

HarmonyOS创造了一个超级虚拟终端互联的世界，将人、设备、场景有机地联系在一起，让消费者通过各种智能终端实现极速发现、极速连接、硬件互助、资源共享等场景体验。一套HarmonyOS系统，能够满足小到耳机，大到车机、智慧屏、手机等设备的需求，让不同设备使用同一语言无缝沟通。

2024年9月，华为正式发布了HarmonyOS NEXT 1.0原生版。根据华为公司发布的数据，截至2024年6月21日，鸿蒙生态设备已经超过9亿台。

项 目 小 结

本项目介绍了Linux、CentOS操作系统和虚拟机，完成了安装VMware Workstation虚拟机软件，创建虚拟机，配置虚拟机网络，使用虚拟机连接远程管理工具、初始化虚拟机等操作，为Docker的安装及相关操作奠定了基础。

习题测试

一、单选题

1. (　　) 是目前全球使用量最多且开源的服务器操作系统。

A. Windows　　B. Mac

C. UNIX　　D. Linux

2. (　　) 更加稳定，开机时间可达几年。

A. Windows　　B. Mac

C. UNIX　　D. Linux

3. 虚拟机是在实体计算机上通过(　　)模拟出的一台或多台虚拟计算机。

A. 硬件　　B. 软件

C. 网络　　D. 虚拟化

二、多选题

1. Linux操作系统的特点有(　　)。

A. 开放性　　B. 多用户

C. 多任务　　D. 稳定性

2. Linux分支有很多，现在比较有名的包括(　　)等。

A. RedHat　　B. Ubuntu

C. Debian　　D. CentOS

3. 国产Linux操作系统有(　　)等。

A. 红旗Linux　　B. 麒麟

C. 深度OS　　D. Oracle Linux

三、简答题

1. 如何创建虚拟机、拍摄快照和克隆？

2. 虚拟机的初始化包括哪些操作？

项目二
安装 Docker 及管理镜像与容器

学习目标

(1) 了解 Docker 容器技术。
(2) 了解计算机虚拟化技术。
(3) 理解 Docker 容器与虚拟机的区别。
(4) 掌握 Docker 的安装。
(5) 掌握 Docke 镜像的基本操作。
(6) 掌握 Docke 容器的基本操作。

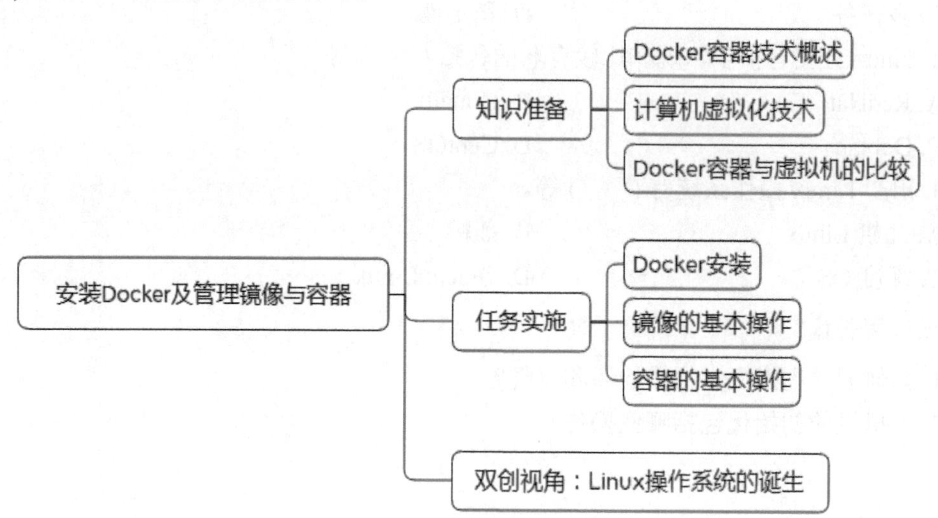

2.1 知识准备

2.1.1 Docker 容器技术概述

云计算是与信息技术、软件、互联网相关的一种服务，所形成的计算资源共享池叫作“云”。云计算把许多计算资源集合起来，通过软件实现自动化管理，只需要很少的人参与，就能快速提供资源给用户。云计算是一种提供资源的网络，使用者可以随时获取“云”上的资源，并按使用量付费，就像自来水厂一样，随时按需使用，根据使用量缴费给自来水厂即可。

云计算的服务类型通常分为三类：基础设施即服务 (IaaS)、平台即服务 (PaaS) 和软件即服务 (SaaS)。基础设施即服务 (IaaS) 是指计算提供商通过网络向用户提供虚拟化计算资源，如虚拟机、存储、网络和操作系统。平台即服务 (PaaS) 是一种服务类别，为开发人员提供通过全球互联网构建的应用程序和服务平台，同时为开发、测试和管理软件应用程序提供按需开发的环境。软件即服务 (SaaS) 是通过互联网提供按需付费的应用程序，云计算供商托管和管理软件应用程序，允许其用户通过网络连接并访问应用程序。

容器是将应用与其所有必要文件捆绑到一个运行环境中的技术。容器可以隔离软件，使其能够在不同的操作系统、硬件、网络、存储系统和安全策略中独立运行。由于操作系统并未打包到容器中，因此每个容器仅需使用极少的计算资源，不仅占用空间极小，而且易于安装。

容器即服务 (CaaS) 是一款云服务，使用户能够管理和部署容器化应用。在云计算服务范畴内，CaaS 被认为是基础架构即服务 (IaaS) 的一种子集，介于 IaaS 和平台即服务 (PaaS) 之间。CaaS 的基本资源为容器，它是云原生应用和微服务的常见部署机制。使用容器有如下优点：

(1) 可移植性好：容器化应用程序无须进行修改或重新编译，便可以在不同的环境中平稳运行。

(2) 可扩展性强：容器具有水平扩展的功能，用户可以在同一集群中成倍增加相同容器的数量，从而满足扩展的需要。由于仅在需要时运行所需容器，因此可以大大降低成本。

(3) 高效性：容器所需的资源要少于虚拟机 (VM)，因为它们不需要单独的操作系统。用户可以在单个服务器上运行多个容器，而且它们需要较少的硬件支持，所以成本更低。

(4) 更高的安全性：容器之间彼此隔离，在一个容器遭到破坏的情况下，其他容器并不会受到影响。

(5) 速度快：由于容器相对于操作系统具有自主性，因此其启动和停止仅需几秒钟的时间。同时容器加快了开发和运维工作，可带来更快、更流畅的用户体验。

2.1.2　计算机虚拟化技术

虚拟化是一个广义的术语，在计算机方面通常是指计算元件在虚拟的基础上而不是真实的基础上运行。虚拟化技术可以扩大硬件的容量，简化软件的重新配置过程。CPU 的虚拟化技术可以单 CPU 模拟多 CPU 并行，允许一个平台同时运行多个操作系统，并且多个应用程序可以在相互独立的空间内运行而互不影响，从而显著提高计算机的工作效率。

虚拟化技术与多任务以及超线程技术是完全不同的。多任务是指在一个操作系统中多个程序同时并行运行；在虚拟化技术中，则可以同时运行多个操作系统，而且每一个操作系统中都有多个程序运行，每一个操作系统都运行在一个虚拟的 CPU 或者是虚拟主机上；而超线程技术只是用单 CPU 模拟双 CPU 来平衡程序运行性能，这两个模拟出来的 CPU 是不能分离的，只能协同工作。

与 VMware Workstation 等同样能达到虚拟效果的软件不同，虚拟化技术是一个巨大的技术进步，具体表现在减少软件虚拟机相关开销和支持更广泛的操作系统等方面。虚拟化技术需要 CPU、主板芯片组、BIOS 和软件等支持，如 VMM(Virtual Machine Monitor，虚拟机监视器) 软件或者某些操作系统本身。即使只是 CPU 支持虚拟化技术，在配合 VMM 软件的情况下，也会比完全不支持虚拟化技术的系统有更好的性能。虚拟化技术具有以下特点：

(1) 保真性：应用系统程序在虚拟机上执行，除了时间因素外 (会比在物理硬件上执行慢一点)，其他方面表现与在物理硬件上执行相同的行为一样。

(2) 高性能：在虚拟环境中，应用程序的绝大多数指令能在虚拟机管理器不受干预的情况下，直接在物理硬件上执行。

(3) 安全性：物理硬件由虚拟机管理器全权管理，虚拟执行环境的程序 (包括操作系统) 不能直接访问物理硬件。

2.1.3　Docker 容器与虚拟机的比较

Docker 作为一种新兴的虚拟化方式，与传统的虚拟化方式相比具有众多的优势。由于容器不需要进行硬件虚拟以及运行完整的操作系统等额外开销，因此 Docker 对系统资源的利用率更高。容器和虚拟机均具有相似的资源隔离和分配优势，它们的不同点在于：每个虚拟机上需要运行一个操作系统，而容器里无须安装操作系统，二者的对比如图 2-1 所示。

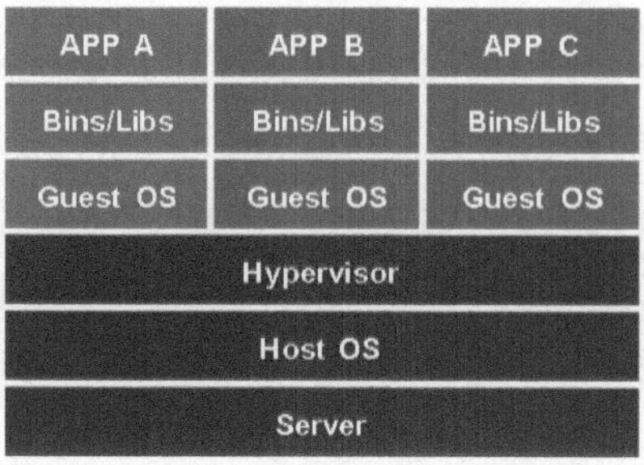

(a) 虚拟机 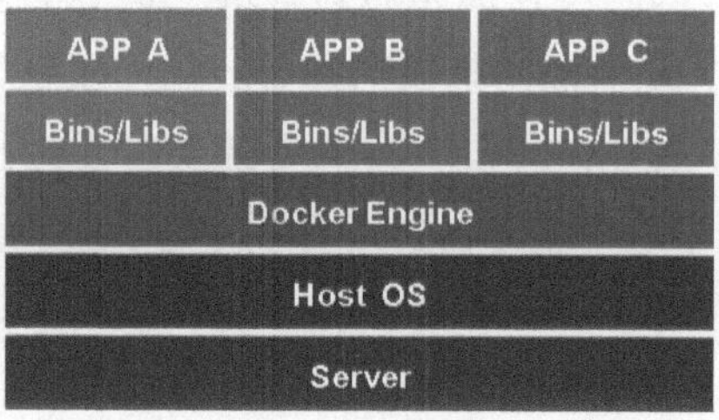　　(b) 容器

图 2-1　虚拟机与容器

将 Docker 容器与虚拟机 (VM) 在操作系统、存储大小、运行性能、移植性、硬件亲和性及部署速度等方面进行比较，不同之处见表 2-1 所示。

表 2-1　Docker 容器与虚拟机 (VM) 的比较

特 点	Docker 容器	虚拟机 (VM)
操作系统	与宿主机共享 OS	宿主机上 OS 运行虚拟机 OS
存储大小	镜像小，便于存储和传输	镜像庞大
运行性能	几乎无额外性能损失	操作系统有额外 CPU 和内存消耗
移植性	灵活、轻松	笨重，与虚拟化技术耦合度高
硬件亲和性	面向软件开发者	面向硬件运维者
部署速度	快速，秒级	较慢，10 s 以上

Docker 容器技术中经常见到一些专有名词，如镜像、镜像仓库、容器、伪终端等，对于它们的解释如下：

镜像：将一个或一组应用以及它们所需的运行环境制作成单一的文件，以方便用户下载和使用。常见的镜像文件格式有 ISO、BIN、IMG、TAO、DAO、CIF、FCD。

存出镜像：当镜像需要在不同的主机上迁移时，首先要将镜像保存为本地文件，此过程称为存出镜像。

载入镜像：将存出的镜像从一台主机复制到另一台主机，需要将该镜像导入到目的主机的镜像库中，此过程称为载入镜像。

镜像仓库：本地存储的镜像很多时，指定的专门存放这些镜像的地方称为镜像仓库。

容器：容器就是镜像运行时的实例，可随时被启动、开始、停止、删除，Docker 利用容器来运行和隔离应用。镜像是只读模板，而容器会增加一个额外的可写层。

导出容器：将一个 Docker 容器从一台主机迁移到另一台主机时，首先需要将已经建立好的容器导出为文件，这个容器处于运行状态或停止状态均可，此过程称为导出容器。

导入容器：导出的文件可以传输到其他主机，在目标机器上通过导入命令实现容器的迁移，此过程称为导入容器。

伪终端：为其他程序提供终端形式的接口，其作用与普通终端相似，区别是伪终端没有对应的硬件设备。

2.2 任务实施

2.2.1 Docker 安装

Docker 安装

1. 任务目标

掌握安装 Docker 的方法。

2. 任务内容

(1) 部署 Docker 安装前的环境。

(2) 配置 docker-ce 源。

(3) 安装 docker-ce。

(4) 配置镜像加速器。

3. 完成任务所需的设备和软件

(1) 一台安装 Windows 10 操作系统的计算机。

(2) VMware Workstation、Docker。

(3) 远程管理工具 MobaXterm。

4. 任务实施步骤

第一步，查看防火墙状态，确认防火墙已关闭，操作命令如下：

```
[root@docker ~]# systemctl status firewalld
```

命令运行结果如图 2-2 所示。

```
[root@docker ~]# systemctl status firewalld
● firewalld.service - firewalld - dynamic firewall daemon
   Loaded: loaded (/usr/lib/systemd/system/firewalld.service; disabled; vendor preset: enabled)
   Active: inactive (dead)
     Docs: man:firewalld(1)
```

图 2-2　查看防火墙状态

如果防火墙状态显示“Active: inactive (dead)”，则表示防火墙关闭；如果防火墙状态显示“Active: active (running)”，则表示防火墙开启，此时需要使用任务 1.2.5 实施步骤中第一步的命令关闭防火墙。

第二步，查看 SELINUX 状态，确保关闭 SELINUX，操作命令如下：

```
[root@docker ~]# cat /etc/selinux/config
```

命令运行结果如图 2-3 所示。

```
[root@docker ~]# cat /etc/selinux/config

# This file controls the state of SELinux on the system.
# SELINUX= can take one of these three values:
#     enforcing - SELinux security policy is enforced.
#     permissive - SELinux prints warnings instead of enforcing.
#     disabled - No SELinux policy is loaded.
SELINUX=disabled
# SELINUXTYPE= can take one of three values:
#     targeted - Targeted processes are protected,
#     minimum - Modification of targeted policy. Only selected processes are protected.
#     mls - Multi Level Security protection.
SELINUXTYPE=targeted
```

图 2-3　查看 SELINUX 状态

如果 SELINUX 的值不是 disabled，需要使用任务 1.2.5 实施步骤中第二步的命令关闭 SELINUX。

第三步，安装所需的依赖包，操作命令如下：

[root@docker ~]# yum install -y yum-utils device-mapper-persistent-data lvm2

命令运行结果如图 2-4 所示。

```
[root@docker ~]# yum install -y yum-utils device-mapper-persistent-data lvm2
Loaded plugins: fastestmirror
Repository cr is listed more than once in the configuration
Repository fasttrack is listed more than once in the configuration
Loading mirror speeds from cached hostfile
Resolving Dependencies
--> Running transaction check
---> Package device-mapper-persistent-data.x86_64 0:0.7.3-3.el7 will be updated
---> Package device-mapper-persistent-data.x86_64 0:0.8.5-3.el7_9.2 will be an update
---> Package lvm2.x86_64 7:2.02.180-8.el7 will be updated
---> Package lvm2.x86_64 7:2.02.187-6.el7_9.5 will be an update
--> Processing Dependency: lvm2-libs = 7:2.02.187-6.el7_9.5 for package: 7:lvm2-2.02.187-6.el7_9.5.x86_64
---> Package yum-utils.noarch 0:1.1.31-54.el7_8 will be installed
--> Processing Dependency: python-kitchen for package: yum-utils-1.1.31-54.el7_8.noarch
--> Running transaction check
---> Package lvm2-libs.x86_64 7:2.02.180-8.el7 will be updated
---> Package lvm2-libs.x86_64 7:2.02.187-6.el7_9.5 will be an update
--> Processing Dependency: device-mapper-event = 7:1.02.170-6.el7_9.5 for package: 7:lvm2-libs-2.02.187-6.el7_9.5.x86_64
---> Package python-kitchen.noarch 0:1.1.1-5.el7 will be installed
--> Processing Dependency: python-chardet for package: python-kitchen-1.1.1-5.el7.noarch
--> Running transaction check
---> Package device-mapper-event.x86_64 7:1.02.149-8.el7 will be updated
---> Package device-mapper-event.x86_64 7:1.02.170-6.el7_9.5 will be an update
--> Processing Dependency: device-mapper-event-libs = 7:1.02.170-6.el7_9.5 for package: 7:device-mapper-event-1.02.170-6.el7_9.5.x86_64
--> Processing Dependency: device-mapper = 7:1.02.170-6.el7_9.5 for package: 7:device-mapper-event-1.02.170-6.el7_9.5.x86_64
---> Package python-chardet.noarch 0:2.2.1-3.el7 will be installed
--> Running transaction check
---> Package device-mapper.x86_64 7:1.02.149-8.el7 will be updated
--> Processing Dependency: device-mapper = 7:1.02.149-8.el7 for package: 7:device-mapper-libs-1.02.149-8.el7.x86_64
---> Package device-mapper.x86_64 7:1.02.170-6.el7_9.5 will be an update
---> Package device-mapper-event-libs.x86_64 7:1.02.149-8.el7 will be updated
---> Package device-mapper-event-libs.x86_64 7:1.02.170-6.el7_9.5 will be an update
--> Running transaction check
---> Package device-mapper-libs.x86_64 7:1.02.149-8.el7 will be updated
---> Package device-mapper-libs.x86_64 7:1.02.170-6.el7_9.5 will be an update
--> Finished Dependency Resolution

Dependencies Resolved

==========================================================================================================
 Package                                  Arch                  Version                       Repository            Size
==========================================================================================================
Installing:
 ......
Complete!
```

图 2-4　安装所需的依赖包

第四步，配置 Docker 的安装源，操作命令如下：

[root@docker ~]# wget -O /etc/yum.repos.d/docker-ce.repo http://mirrors.aliyun.com/docker-ce/linux/centos/docker-ce.repo

命令运行结果如图 2-5 所示。

```
[root@docker ~]# wget -O /etc/yum.repos.d/docker-ce.repo http://mirrors.aliyun.com/docker-ce/linux/centos/docker-ce.repo
--2024-08-25 09:14:07--  http://mirrors.aliyun.com/docker-ce/linux/centos/docker-ce.repo
Resolving mirrors.aliyun.com (mirrors.aliyun.com)... 182.40.42.230, 182.40.42.231, 182.40.42.232
Connecting to mirrors.aliyun.com (mirrors.aliyun.com)|182.40.42.230|:80... connected.
HTTP request sent, awaiting response... 200 OK
Length: 2081 (2.0K) [application/octet-stream]
Saving to: '/etc/yum.repos.d/docker-ce.repo'

100%[=====================================================================================>] 2,081       --.-K/s   in 0s

2024-08-25 09:14:07 (272 MB/s) - '/etc/yum.repos.d/docker-ce.repo' saved [2081/2081]

[root@docker ~]# 
```

图 2-5　配置 Docker 的安装源

第五步，安装 Docker，操作命令如下：

[root@docker ~]# yum install docker-ce -y

命令运行结果如图 2-6 所示。

```
[root@docker ~]# yum install docker-ce -y
Loaded plugins: fastestmirror
Repository cr is listed more than once in the configuration
Repository fasttrack is listed more than once in the configuration
Loading mirror speeds from cached hostfile
docker-ce-stable                                              | 3.5 kB  00:00:00
epel                                                          | 4.3 kB  00:00:00
extras                                                        | 2.9 kB  00:00:00
os                                                            | 3.6 kB  00:00:00
updates                                                       | 2.9 kB  00:00:00
(1/2): docker-ce-stable/7/x86_64/primary_db                   | 152 kB  00:00:00
(2/2): docker-ce-stable/7/x86_64/updateinfo                   |   55 B  00:00:00
Resolving Dependencies
--> Running transaction check
---> Package docker-ce.x86_64 3:26.1.4-1.el7 will be installed
--> Processing Dependency: container-selinux >= 2:2.74 for package: 3:docker-ce-26.1.4-1.el7.x86_64
--> Processing Dependency: containerd.io >= 1.6.24 for package: 3:docker-ce-26.1.4-1.el7.x86_64
--> Processing Dependency: libseccomp >= 2.3 for package: 3:docker-ce-26.1.4-1.el7.x86_64
--> Processing Dependency: docker-ce-cli for package: 3:docker-ce-26.1.4-1.el7.x86_64
--> Processing Dependency: docker-ce-rootless-extras for package: 3:docker-ce-26.1.4-1.el7.x86_64
--> Processing Dependency: libcgroup for package: 3:docker-ce-26.1.4-1.el7.x86_64
--> Running transaction check
---> Package container-selinux.noarch 2:2.119.2-1.911c772.el7_8 will be installed
--> Processing Dependency: policycoreutils-python for package: 2:container-selinux-2.119.2-1.911c772.el7_8.noarch
---> Package containerd.io.x86_64 0:1.6.33-3.1.el7 will be installed
---> Package docker-ce-cli.x86_64 1:26.1.4-1.el7 will be installed
--> Processing Dependency: docker-buildx-plugin for package: 1:docker-ce-cli-26.1.4-1.el7.x86_64
--> Processing Dependency: docker-compose-plugin for package: 1:docker-ce-cli-26.1.4-1.el7.x86_64
---> Package docker-ce-rootless-extras.x86_64 0:26.1.4-1.el7 will be installed
--> Processing Dependency: fuse-overlayfs >= 0.7 for package: docker-ce-rootless-extras-26.1.4-1.el7.x86_64
--> Processing Dependency: slirp4netns >= 0.4 for package: docker-ce-rootless-extras-26.1.4-1.el7.x86_64
---> Package libcgroup.x86_64 0:0.41-21.el7 will be installed
---> Package libseccomp.x86_64 0:2.3.1-4.el7 will be installed
--> Running transaction check
---> Package docker-buildx-plugin.x86_64 0:0.14.1-1.el7 will be installed
---> Package docker-compose-plugin.x86_64 0:2.27.1-1.el7 will be installed
---> Package fuse-overlayfs.x86_64 0:0.7.2-6.el7_8 will be installed
--> Processing Dependency: libfuse3.so.3(FUSE_3.2)(64bit) for package: fuse-overlayfs-0.7.2-6.el7_8.x86_64
--> Processing Dependency: libfuse3.so.3(FUSE_3.0)(64bit) for package: fuse-overlayfs-0.7.2-6.el7_8.x86_64
--> Processing Dependency: libfuse3.so.3()(64bit) for package: fuse-overlayfs-0.7.2-6.el7_8.x86_64
---> Package policycoreutils-python.x86_64 0:2.5-34.el7 will be installed
--> Processing Dependency: policycoreutils = 2.5-34.el7 for package: policycoreutils-python-2.5-34.el7.x86_64
--> Processing Dependency: setools-libs >= 3.3.8-4 for package: policycoreutils-python-2.5-34.el7.x86_64
......
Complete!
[root@docker ~]#
```

图 2-6 安装 Docker

第六步，启动 Docker，并将其设置为开机自启动，操作命令如下：

```
[root@docker ~]# systemctl start docker
[root@docker ~]# systemctl enable docker
```

命令运行结果如图 2-7 所示。

```
[root@docker ~]# systemctl start docker
[root@docker ~]# systemctl enable docker
Created symlink from /etc/systemd/system/multi-user.target.wants/docker.service to /usr/lib/systemd/system/docker.service.
[root@docker ~]#
```

图 2-7 启动 Docker，并将其设置为开机自启动

第七步，查看 Docker 版本号，操作命令如下：

```
[root@docker ~]# docker --version
```

命令运行结果如图 2-8 所示。

```
[root@docker ~]# docker --version
Docker version 26.1.4, build 5650f9b
[root@docker ~]#
```

图 2-8 查看 Docker 版本号

第八步，查看 Docker 信息，操作命令如下：

```
[root@docker ~]# docker info
```

命令运行结果如图 2-9 所示。

```
[root@docker ~]# docker info
Client: Docker Engine - Community
 Version:    26.1.4
 Context:    default
 Debug Mode: false
 Plugins:
  buildx: Docker Buildx (Docker Inc.)
    Version:  v0.14.1
    Path:     /usr/libexec/docker/cli-plugins/docker-buildx
  compose: Docker Compose (Docker Inc.)
    Version:  v2.27.1
    Path:     /usr/libexec/docker/cli-plugins/docker-compose

Server:
 Containers: 0
  Running: 0
  Paused: 0
  Stopped: 0
 Images: 0
 Server Version: 26.1.4
 Storage Driver: overlay2
  Backing Filesystem: xfs
  Supports d_type: true
  Using metacopy: false
  Native Overlay Diff: true
  userxattr: false
 Logging Driver: json-file
 Cgroup Driver: cgroupfs
 Cgroup Version: 1
 Plugins:
  Volume: local
  Network: bridge host ipvlan macvlan null overlay
  Log: awslogs fluentd gcplogs gelf journald json-file local splunk syslog
 Swarm: inactive
 Runtimes: io.containerd.runc.v2 runc
 Default Runtime: runc
 Init Binary: docker-init
 containerd version: d2d58213f83a351ca8f528a95fbd145f5654e957
 runc version: v1.1.12-0-g51d5e94
 init version: de40ad0
 Security Options:
  seccomp
   Profile: builtin
```

图 2-9　查看 Docker 信息

第九步，配置 Docker 镜像加速器，便于 Docker 更快地拉取镜像，操作命令如下：

```
[root@docker ~]# vim /etc/docker/daemon.json
{
  "registry-mirrors": ["http://hubmirror.c.163.com"]
}
[root@docker ~]# systemctl daemon-reload
[root@docker ~]# systemctl restart docker
```

这里配置了网易云镜像加速器，如果使用阿里云镜像加速器，每个用户就会有专属的镜像加速地址，获取该地址的方法如下：

(1) 访问阿里云网站 (https://www.aliyun.com)，注册账号并登录。

(2) 从“产品”→“容器”→“容器服务”→“容器镜像服务ACR(Alibaba cloud Container Registry)”→“管理控制台”→“镜像工具”→“镜像加速器”的路径找到镜像加速器，即可获取个人专属的镜像加速地址，如图2-10所示。

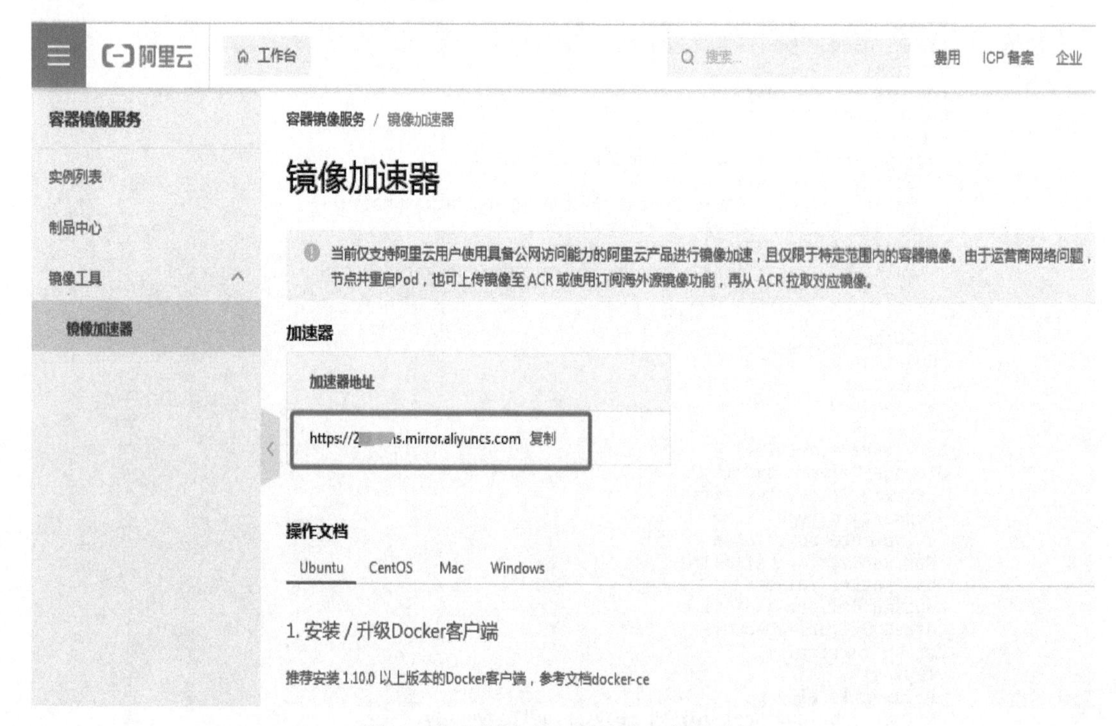

图2-10 获得个人专属的镜像加速地址

注意：如果以上两个镜像加速器都不奏效，建议配置以下镜像加速器。

```
[root@docker ~]# vim /etc/docker/daemon.json
{
"registry-mirrors":
 [
  "https://docker.mirrors.sjtug.sjtu.edu.cn",
  "https://docker.m.daocloud.io",
  "https://noohub.ru",
  "https://huecker.io",
  "https://dockerhub.timeweb.cloud",
  "https://registry.cn-hangzhou.aliyuncs.com"
 ]
}
[root@docker ~]# systemctl daemon-reload
[root@docker ~]# systemctl restart docker
```

第十步，为虚拟机拍摄快照，然后保存虚拟机此时的状态，如图2-11所示。

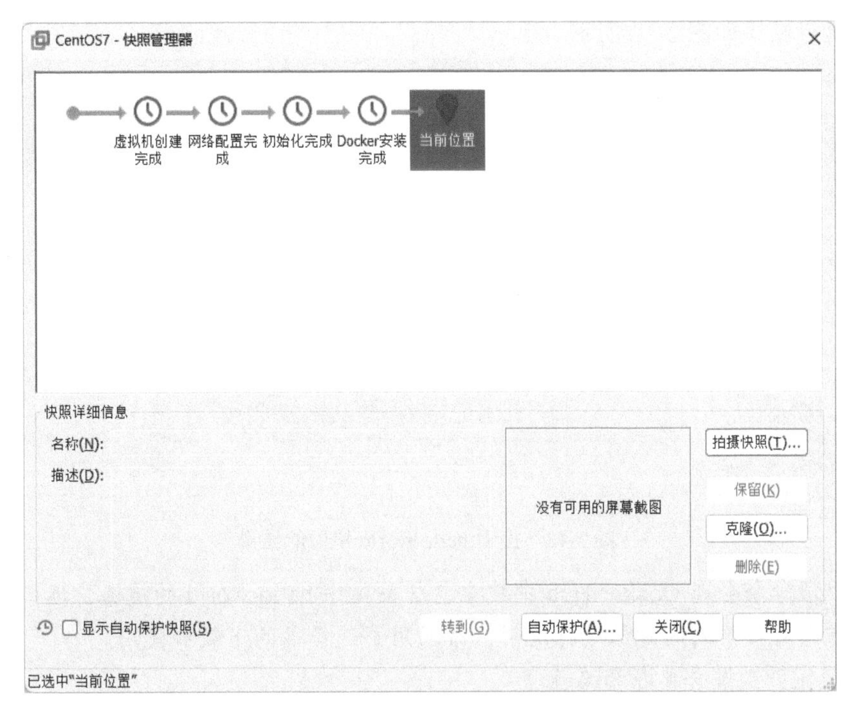

图 2-11　拍摄快照“Docker 安装完成”

第十一步，对此虚拟机进行克隆，克隆后的虚拟机命名为 CentOS7-m，将其作为母机以便快速生成多台同样配置的虚拟机。

2.2.2　镜像的基本操作

镜像的基本操作

1. 任务目标

掌握镜像的基本操作方法。

2. 任务内容

(1) 搜索、拉取镜像。

(2) 查看、修改镜像信息。

(3) 删除镜像。

(4) 迁移镜像。

3. 完成任务所需的设备和软件

(1) 一台安装 Windows 10 操作系统的计算机。

(2) VMware Workstation、Docker。

(3) 远程管理工具 MobaXterm。

4. 任务实施步骤

第一步，搜索镜像，如搜索 hello-world 镜像，操作命令如下：

```
[root@docker ~]# docker search hello-world
```

命令运行结果如图 2-12 所示。

```
[root@docker ~]# docker search hello-world
NAME                                       DESCRIPTION                                       STARS     OFFICIAL   AUTOMATED
hello-world                                Hello World! (an example of minimal Dockeriz…   1788      [OK]
kitematic/hello-world-nginx                A light-weight nginx container that demonstr…   152
tutum/hello-world                          Image to test docker deployments. Has Apache…   89                   [OK]
dockercloud/hello-world                    Hello World!                                    19                   [OK]
crccheck/hello-world                       Hello World web server in under 2.5 MB          15                   [OK]
ansibleplaybookbundle/hello-world-db-apb   An APB which deploys a sample Hello World! a…   2                    [OK]
ppc64le/hello-world                        Hello World! (an example of minimal Dockeriz…   2
datawire/hello-world                       Hello World! Simple Hello World implementati…   1                    [OK]
ansibleplaybookbundle/hello-world-apb      An APB which deploys a sample Hello World! a…   1                    [OK]
rancher/hello-world                                                                        1
thomaspoignant/hello-world-rest-json       This project is a REST hello-world API to bu…   1
koudaiii/hello-world                                                                       0
strimzi/hello-world-producer                                                               0
strimzi/hello-world-consumer                                                               0
businessgeeks00/hello-world-nodejs                                                         0
strimzi/hello-world-streams                                                                0
freddiedevops/hello-world-spring-boot                                                      0
tsepotesting123/hello-world                                                                0
armswdev/c-hello-world                     Simple hello-world C program on Alpine Linux…   0
garystafford/hello-world                   Simple hello-world Spring Boot service for t…   0                    [OK]
dandando/hello-world-dotnet                                                                0
rsperling/hello-world3                                                                     0
kevindockercompany/hello-world                                                             0
okteto/hello-world                                                                         0
uniplaces/hello-world                                                                      0
```

图 2-12 搜索 hello-world 镜像的结果

注意：该命令会在 Docker Hub 中搜索包含关键字 hello-world 的镜像，返回信息如下，通过这些信息对搜索到的镜像进行说明，帮助用户选择性的下载和使用。

- NAME：镜像仓库源的名称。
- DESCRIPTION：镜像的描述。
- OFFICIAL：是否 docker 官方发布。
- STARS：受欢迎程度。
- AUTOMATED：是否自动创建。

第二步，从 Docker Hub 拉取镜像到本地，如拉取 hello-world 镜像，操作命令如下：

```
[root@docker ~]# docker pull hello-world
```

命令运行结果如图 2-13 所示。

```
[root@docker ~]# docker pull hello-world
Using default tag: latest
latest: Pulling from library/hello-world
2db29710123e: Pull complete
Digest: sha256:13e367d31ae85359f42d637adf6da428f76d75dc9afeb3c21faea0d976f5c651
Status: Downloaded newer image for hello-world:latest
docker.io/library/hello-world:latest
```

图 2-13 拉取 hello-world 镜像到本地

返回信息如下：

- Using default tag：latest 表示使用默认标签，“latest”表示最新版本。
- Latest：Pulling from library/hello-world 表示正在从官方仓库 library/hello-world 拉取最新版的 hello-world 镜像。
- 2db29710123e：Pull complete 表示镜像的一个层已经被拉取。
- Digest：sha256：13e…表示镜像摘要，包括了镜像的所有层的哈希值，用于验证镜像内容的完整性。
- Status：Downloaded newer image for hello-world：latest 表示新的镜像 hello-world 已被下载。

• docker.io/library/ hello-world：latest 表示完整的镜像名，包括仓库地址和标签。

第三步，查看本地所有镜像，操作命令如下：

```
[root@docker ~]# docker images
```

命令运行结果如图 2-14 所示。

```
[root@docker ~]# docker images
REPOSITORY              TAG       IMAGE ID       CREATED         SIZE
compose_lnmp_nginx      latest    4bd05f929494   2 months ago    497MB
compose_lnmp_php        latest    cfeec644cc0b   2 months ago    1.19GB
centos                  mysql     7df9170a1a89   3 months ago    476MB
tomcat                  centos    40552ff22f8f   3 months ago    626MB
nginx                   new       b12a1a921e1f   3 months ago    471MB
local/c7-systemd        latest    12ee522a64b2   3 months ago    365MB
sshd                    new       a7742ca6b111   3 months ago    365MB
my_tomcat               v1        70dafa8f2f2a   3 months ago    612MB
mysql                   5.7       f26e21ddd20d   3 months ago    450MB
httpd                   latest    118b6abfbf55   3 months ago    144MB
mysql                   5.6       dd3b2a5dcb48   6 months ago    303MB
hello-world             latest    feb5d9fea6a5   9 months ago    13.3kB
centos                  7         eeb6ee3f44bd   9 months ago    204MB
```

图 2-14　查看本地镜像

返回信息如下：

• REPOSITORY：镜像的仓库源。

• TAG：镜像的标签，默认是 latest。

• IMAGE ID：镜像的 ID 号，是唯一的标识符。

• CREATED：镜像的创建时间。

• SIZE：镜像大小。

第四步，查看本地镜像 hello-world 的详细信息，操作命令如下：

```
[root@docker ~]# docker inspect hello-world
```

第五步：修改本地镜像 hello-world 的名称为 my-images，标签改为 v1.0，操作命令如下：

```
[root@docker ~]# docker tag hello-world:latest my-images:v1.0
```

命令运行之后，查看本地镜像如图 2-15 所示。

```
[root@docker ~]# docker tag hello-world:latest my-images:v1.0
[root@docker ~]# docker images
REPOSITORY              TAG       IMAGE ID       CREATED         SIZE
compose_lnmp_nginx      latest    4bd05f929494   2 months ago    497MB
compose_lnmp_php        latest    cfeec644cc0b   2 months ago    1.19GB
centos                  mysql     7df9170a1a89   3 months ago    476MB
tomcat                  centos    40552ff22f8f   3 months ago    626MB
nginx                   new       b12a1a921e1f   3 months ago    471MB
local/c7-systemd        latest    12ee522a64b2   3 months ago    365MB
sshd                    new       a7742ca6b111   3 months ago    365MB
my_tomcat               v1        70dafa8f2f2a   3 months ago    612MB
mysql                   5.7       f26e21ddd20d   3 months ago    450MB
httpd                   latest    118b6abfbf55   3 months ago    144MB
mysql                   5.6       dd3b2a5dcb48   6 months ago    303MB
hello-world             latest    feb5d9fea6a5   9 months ago    13.3kB
my-images               v1.0      feb5d9fea6a5   9 months ago    13.3kB
centos                  7         eeb6ee3f44bd   9 months ago    204MB
```

图 2-15　修改本地镜像名称和标签

注意：可以仅修改镜像的名称或标签，如仅修改 hello-world 镜像标签的操作命令如下所示。

```
[root@docker ~]# docker tag hello-world:latest hello-world:v1.0
```

命令运行之后，查看本地镜像如图 2-16 所示。值得一提的是，修改名称或标签之后的镜像与修改之前的镜像 ID 相同。

```
[root@docker ~]# docker tag hello-world:latest hello-world:v1.0
[root@docker ~]# docker images
REPOSITORY                TAG       IMAGE ID       CREATED        SIZE
compose_lnmp_nginx        latest    4bd05f929494   2 months ago   497MB
compose_lnmp_php          latest    cfeec644cc0b   2 months ago   1.19GB
centos                    mysql     7df9170a1a89   3 months ago   476MB
tomcat                    centos    40552ff22f8f   3 months ago   626MB
nginx                     new       b12a1a921e1f   3 months ago   471MB
local/c7-systemd          latest    12ee522a64b2   3 months ago   365MB
sshd                      new       a7742ca6b111   3 months ago   365MB
my_tomcat                 v1        70dafa8f2f2a   3 months ago   612MB
mysql                     5.7       f26e21ddd20d   3 months ago   450MB
httpd                     latest    118b6abfbf55   3 months ago   144MB
mysql                     5.6       dd3b2a5dcb48   6 months ago   303MB
my-images                 v1.0      feb5d9fea6a5   9 months ago   13.3kB
hello-world               latest    feb5d9fea6a5   9 months ago   13.3kB
hello-world               v1.0      feb5d9fea6a5   9 months ago   13.3kB
centos                    7         eeb6ee3f44bd   9 months ago   204MB
```

图 2-16　修改本地镜像标签

第六步，删除本地镜像 my-images，操作命令如下：

```
[root@docker ~]# docker rmi my-images:v1.0
```

命令运行之后，查看本地镜像如图 2-17 所示。

```
[root@docker ~]# docker rmi my-images:v1.0
Untagged: my-images:v1.0
[root@docker ~]# docker images
REPOSITORY                TAG       IMAGE ID       CREATED        SIZE
compose_lnmp_nginx        latest    4bd05f929494   2 months ago   497MB
compose_lnmp_php          latest    cfeec644cc0b   2 months ago   1.19GB
centos                    mysql     7df9170a1a89   3 months ago   476MB
tomcat                    centos    40552ff22f8f   3 months ago   626MB
nginx                     new       b12a1a921e1f   3 months ago   471MB
local/c7-systemd          latest    12ee522a64b2   3 months ago   365MB
sshd                      new       a7742ca6b111   3 months ago   365MB
my_tomcat                 v1        70dafa8f2f2a   3 months ago   612MB
mysql                     5.7       f26e21ddd20d   3 months ago   450MB
httpd                     latest    118b6abfbf55   3 months ago   144MB
mysql                     5.6       dd3b2a5dcb48   6 months ago   303MB
hello-world               latest    feb5d9fea6a5   9 months ago   13.3kB
hello-world               v1.0      feb5d9fea6a5   9 months ago   13.3kB
centos                    7         eeb6ee3f44bd   9 months ago   204MB
```

图 2-17　删除本地镜像 my-images:v1.0

第七步，镜像迁移，进行此操作之前，需先拉取镜像 centos/httpd，然后利用母机 CentOS7-m 克隆另外一台虚拟机 Client(IP 地址为 192.168.1.100)，最后实现在两台虚拟机之间的镜像迁移，操作步骤如下：

(1) 在 Docker 主机上，存出本地镜像 centos/httpd 为文件 httpd，操作命令如下：

```
[root@docker ~]# ls
[root@docker ~]# docker save -o httpd centos/httpd
[root@docker ~]# ls
```

命令运行结果如图 2-18 所示，可见 httpd 文件已经生成。

```
[root@docker ~]# ls
anaconda-ks.cfg  compose_lnmp  constar.tar  consul  myh-w  nginx  sshd  systemctl  tomcat
[root@docker ~]# docker save -o httpd centos/httpd
[root@docker ~]# ls
anaconda-ks.cfg  compose_lnmp  constar.tar  consul  httpd  myh-w  nginx  sshd  systemctl  tomcat
```

图 2-18　查看当前目录文件

(2) 远程复制文件 httpd 到 Client 主机，操作命令如下：

```
[root@docker ~]# scp httpd root@192.168.1.100:httpd
```

命令运行结果如图 2-19 所示，输入 Client 主机的 root 用户密码并回传，结果如图 2-20 所示，表明远程复制成功。

```
[root@docker ~]# scp httpd root@192.168.1.100:httpd
root@192.168.1.100's password:
```

图 2-19　远程复制文件 httpd 到 Client 主机

```
[root@docker ~]# scp httpd root@192.168.1.100:httpd
root@192.168.1.100's password:
httpd
```

图 2-20　远程复制文件完成

(3) 在 Client 主机上，查看当期目录情况，如图 2-21 所示，可见 httpd 文件已经存在。

```
[root@client ~]# ls
anaconda-ks.cfg  compose_lnmp  constar.tar  consul  myh-w  mysshd1  nginx  sshd  systemctl  tomcat
[root@client ~]# ls
anaconda-ks.cfg  compose_lnmp  constar.tar  consul  httpd  myh-w  mysshd1  nginx  sshd  systemctl  tomcat
[root@client ~]#
```

图 2-21　远程复制命令执行前后的当期目录情况

(4) 将文件 httpd 载入到本地镜像库中，操作命令如下：

```
[root@client~]# docker load -i httpd( 或 docker load <httpd)
[root@client~]# docker images
```

命令运行结果如图 2-22 所示。

```
[root@client ~]# docker images
REPOSITORY   TAG       IMAGE ID   CREATED   SIZE
[root@client ~]# docker load -i httpd
071d8bd76517: Loading layer [==================================================>]  210.2MB/210.2MB
d15c61d3ecda: Loading layer [==================================================>]  56.88MB/56.88MB
7c937d8a9f4f: Loading layer [==================================================>]  2.048kB/2.048kB
920640105caf: Loading layer [==================================================>]  2.048kB/2.048kB
Loaded image: centos/httpd:latest
[root@client ~]# docker images
REPOSITORY     TAG      IMAGE ID       CREATED       SIZE
centos/httpd   latest   2cc07fbb5000   5 years ago   258MB
[root@client ~]#
```

图 2-22　将文件 httpd 载入到本地镜像库中

2.2.3　容器的基本操作

容器的基本操作

1. 任务目标

掌握容器的基本操作方法。

2. 任务内容

(1) 创建与启动容器。

(2) 运行与终止容器。

(3) 进入与退出容器。

(4) 删除容器。

(5) 迁移容器。

3. 完成任务所需的设备和软件

(1) 一台安装 Windows 10 操作系统的计算机。

(2) VMware Workstation、Docker。

(3) 远程管理工具 MobaXterm。

4. 任务实施步骤

第一步，创建容器，但不启动容器，操作命令如下：

```
[root@docker ~]# docker create -it centos/httpd /bin/bash
```

-i：让容器的标准输入保持打开状态。

-t：让 Docker 分配一个伪终端。

/bin/bash：用于启动一个交互式的 Bash shell(与操作系统进行交互的一种命令行接口)，以便用户进入容器的命令行界面进行各种管理操作。

命令运行结果返回一个容器 ID，如图 2-23 所示。

```
[root@docker ~]# docker create -it centos/httpd /bin/bash
e2a545fb7f049e88874dbd73a6e4075d5550c46f1bae86dd936283fdc838f853
[root@docker ~]# 
```

图 2-23　创建容器

第二步，查看所有容器及其运行状态，操作命令如下：

```
[root@docker ~]# docker ps -a
```

命令运行结果如图 2-24 所示。

```
[root@docker ~]# docker ps -a
CONTAINER ID   IMAGE          COMMAND       CREATED          STATUS    PORTS     NAMES
17c462f35d61   centos/httpd   "/bin/bash"   7 seconds ago    Created             gifted_darwin
[root@docker ~]# 
```

图 2-24　查看所有容器及其运行状态

返回信息如下：

- CONTAINER ID：容器的 ID 号。
- IMAGE：加载镜像。
- COMMAND：容器内运行的命令
- CREATED：容器创建的时间。
- STATUS：容器的状态。
- PORTS：端口映射。
- NAMES：容器的名称。当没有指定容器名称时，系统会随机分配。

第三步，创建名称为 conht 的容器，操作命令如下：

```
[root@docker ~]# docker create -it --name conht centos/httpd /bin/bash
[root@docker ~]# docker ps -a
```

命令运行结果如图 2-25 所示。

```
[root@docker ~]# docker create -it --name conht centos/httpd /bin/bash
9cea19e0af9823be5f3d3cff20e1229f82bcb8573d64239bbe0f340b7ada8049
[root@docker ~]# docker ps -a
CONTAINER ID   IMAGE          COMMAND       CREATED          STATUS    PORTS     NAMES
9cea19e0af98   centos/httpd   "/bin/bash"   12 seconds ago   Created             conht
17c462f35d61   centos/httpd   "/bin/bash"   18 minutes ago   Created             gifted_darwin
[root@docker ~]# 
```

图 2-25　按照指定名称创建容器

第四步，启动容器 conht，操作命令如下：

[root@docker ~]# docker start conht(或为容器 ID 号)

[root@docker ~]# docker ps -a

命令运行结果如图 2-26 所示。

```
[root@docker ~]# docker start conht
conht
[root@docker ~]# docker ps -a
CONTAINER ID   IMAGE          COMMAND       CREATED          STATUS         PORTS    NAMES
9cea19e0af98   centos/httpd   "/bin/bash"   6 minutes ago    Up 6 seconds   80/tcp   conht
17c462f35d61   centos/httpd   "/bin/bash"   24 minutes ago   Created                 gifted_darwin
[root@docker ~]#
```

图 2-26　启动容器

第五步，终止容器 conht 运行，操作命令如下：

[root@docker ~]# docker stop conht(或为容器 ID 号)

[root@docker ~]# docker ps -a

命令运行结果如图 2-27 所示。

```
[root@docker ~]# docker stop conht
conht
[root@docker ~]# docker ps -a
CONTAINER ID   IMAGE          COMMAND       CREATED          STATUS                       PORTS   NAMES
9cea19e0af98   centos/httpd   "/bin/bash"   11 minutes ago   Exited (137) 5 seconds ago           conht
17c462f35d61   centos/httpd   "/bin/bash"   29 minutes ago   Created                              gifted_darwin
[root@docker ~]#
```

图 2-27　终止容器运行

第六步，创建并启动容器，相当于先创建再启动容器，并执行命令“ls”，操作命令如下：

[root@docker ~]# docker run centos/httpd /bin/bash -c ls

[root@docker ~]# docker ps -a

命令运行结果如图 2-28 所示。若容器中的 shell 命令运行，则容器运行，若 shell 命令结束，则容器退出。

```
[root@docker ~]# docker run centos/httpd /bin/bash -c ls
anaconda-post.log
bin
boot
dev
etc
home
lib
lib64
media
mnt
opt
proc
root
run
run-httpd.sh
sbin
srv
sys
tmp
usr
var
[root@docker ~]# docker ps -a
CONTAINER ID   IMAGE          COMMAND             CREATED          STATUS                        PORTS   NAMES
cebc7081f77b   centos/httpd   "/bin/bash -c ls"   8 minutes ago    Exited (0) 8 minutes ago              unruffled_bardeen
9cea19e0af98   centos/httpd   "/bin/bash"         23 minutes ago   Exited (137) 12 minutes ago           conht
17c462f35d61   centos/httpd   "/bin/bash"         41 minutes ago   Created                               gifted_darwin
[root@docker ~]#
```

图 2-28　创建并启动容器

第七步，利用镜像 centos/httpd 持续运行一个容器，即 Docker 容器以守护进程的形式在后台运行，此时要求容器所运行的程序不能结束。操作命令如下：

```
[root@docker ~]# docker run -d centos/httpd /bin/bash -c "while true;do echo Welcome;done"
[root@docker ~]# docker ps -a
```

命令运行结果如图 2-29 所示。

```
[root@docker ~]# docker run -d centos/httpd /bin/bash -c "while true;do echo Welcome;done"
9e24e0c6c37342594412e9d200742c56cd76da9f68f2c969103f015c720d1928
[root@docker ~]# docker ps -a
CONTAINER ID   IMAGE          COMMAND                  CREATED              STATUS                        PORTS    NAMES
9e24e0c6c373   centos/httpd   "/bin/bash -c 'while…"   6 seconds ago        Up 4 seconds                  80/tcp   hardcore_chandrasekhar
cebc7081f77b   centos/httpd   "/bin/bash -c ls"        39 minutes ago       Exited (0) 39 minutes ago              unruffled_bardeen
9cea19e0af98   centos/httpd   "/bin/bash"              54 minutes ago       Exited (137) 43 minutes ago            conht
17c462f35d61   centos/httpd   "/bin/bash"              About an hour ago    Created                                gifted_darwin
[root@docker ~]#
```

图 2-29　持续运行一个容器

第八步，进入第七步生成的且正在运行的容器，在容器中运行 ls 命令，并退出容器，操作命令如下：

```
[root@docker ~]# docker exec -it 9e24e0c6c373 /bin/bash
[root@9e24e0c6c373 /]# ls
[root@9e24e0c6c373 /]# exit
```

命令运行结果如图 2-30 所示。

```
[root@docker ~]# docker exec -it 9e24e0c6c373 /bin/bash
[root@9e24e0c6c373 /]# ls
anaconda-post.log  bin  boot  dev  etc  home  lib  lib64  media  mnt  opt  proc  root  run  run-httpd.sh  sbin  srv  sys  tmp  usr  var
[root@9e24e0c6c373 /]# exit
exit
[root@docker ~]#
```

图 2-30　进入容器，运行 ls 命令并退出

第九步，查看第七步生成容器内部的输出内容，操作命令如下：

```
[root@docker ~]# docker logs 9e24e0c6c373
```

命令运行结果如图 2-31 所示，可以看到正在持续输出“Welcome”。

```
[root@docker ~]# docker logs 9e24e0c6c373
Welcome
Welcome
Welcome
Welcome
Welcome
Welcome
Welcome
Welcome
Welcome
Welcome
Welcome
Welcome
Welcome
Welcome
Welcome
Welcome
Welcome
Welcome
Welcome
Welcome
Welcome
Welcome
```

图 2-31　查看容器内部的输出内容

第十步，删除处于创建或终止状态的容器，操作命令如下：

```
[root@docker ~]# docker rm 17c462f35d61( 或为容器名称 )
```

[root@docker ~]# docker rm 9cea19e0af98 cebc7081f77b

[root@docker ~]# docker ps -a

命令运行结果如图 2-32 所示。

```
[root@docker ~]# docker ps -a
CONTAINER ID   IMAGE          COMMAND                  CREATED            STATUS                            PORTS     NAMES
9e24e0c6c373   centos/httpd   "/bin/bash -c 'while…"   24 minutes ago     Up 24 minutes                     80/tcp    hardcore_chandrasekhar
cebc7081f77b   centos/httpd   "/bin/bash -c ls"        About an hour ago  Exited (0) About an hour ago                unruffled_bardeen
9cea19e0af98   centos/httpd   "/bin/bash"              About an hour ago  Exited (137) About an hour ago              conht
17c462f35d61   centos/httpd   "/bin/bash"              2 hours ago        Created                                     gifted_darwin
[root@docker ~]# docker rm 17c462f35d61
17c462f35d61
[root@docker ~]# docker rm 9cea19e0af98 cebc7081f77b
9cea19e0af98
cebc7081f77b
[root@docker ~]# docker ps -a
CONTAINER ID   IMAGE          COMMAND                  CREATED          STATUS          PORTS     NAMES
9e24e0c6c373   centos/httpd   "/bin/bash -c 'while…"   26 minutes ago   Up 26 minutes   80/tcp    hardcore_chandrasekhar
[root@docker ~]#
```

图 2-32　删除处于创建或终止状态的容器

第十一步，强制删除正在运行的容器，操作命令如下：

[root@docker ~]# docker rm -f 9e24e0c6c373

[root@docker ~]# docker ps -a

命令运行结果如图 2-33 所示。对于正在运行的容器，建议先终止运行然后再删除。

```
[root@docker ~]# docker rm -f 9e24e0c6c373
9e24e0c6c373
[root@docker ~]# docker ps -a
CONTAINER ID   IMAGE     COMMAND   CREATED   STATUS    PORTS     NAMES
[root@docker ~]#
```

图 2-33　强制删除正在运行的容器

第十二步，容器迁移，实现 Docker 主机和 Client 主机之间的容器迁移，操作步骤如下：

(1) 在 Docker 主机上运行一个容器，操作命令如下：

[root@docker ~]# docker run centos/httpd /bin/bash

[root@docker ~]# docker ps -a

命令运行结果如图 2-34 所示。

```
[root@docker ~]# docker run centos/httpd /bin/bash
[root@docker ~]# docker ps -a
CONTAINER ID   IMAGE          COMMAND       CREATED         STATUS                     PORTS     NAMES
7badee46a197   centos/httpd   "/bin/bash"   8 seconds ago   Exited (0) 7 seconds ago             objective_joliot
[root@docker ~]#
```

图 2-34　在 Docker 主机上运行一个容器

(2) 将容器 7badee46a197 导出为文件 myhttpd，操作命令如下：

[root@docker ~]# ls

[root@docker ~]# docker export 7badee46a197 >myhttpd

[root@docker ~]# ls

命令运行结果如图 2-35 所示，可见文件 myhttpd 已经生成。

```
[root@docker ~]# ls
anaconda-ks.cfg  compose_lnmp  constar.tar  consul  httpd  myh-w  mysshd1  nginx  sshd  systemctl  tomcat
[root@docker ~]# docker export 7badee46a197 >myhttpd
[root@docker ~]# ls
anaconda-ks.cfg  compose_lnmp  constar.tar  consul  httpd  myhttpd  myh-w  mysshd1  nginx  sshd  systemctl  tomcat
[root@docker ~]#
```

图 2-35　将容器导出为文件

(3) 将文件 myhttpd 远程复制到 Client 主机，输入 Client 主机的 root 用户和密码并点击回车键，操作命令如下：

```
[root@docker ~]# scp myhttpd root@192.168.1.100:myhttpd
```

命令运行结果如图 2-36 所示。

```
[root@docker ~]# scp myhttpd root@192.168.1.100:myhttpd
root@192.168.1.100's password:
myhttpd
[root@docker ~]#
```

图 2-36　将文件 myhttpd 远程复制到 Client 主机

(4) 在 Client 主机上，查看当期目录情况，如图 2-37 所示，可见 myhttpd 文件已经存在。

```
[root@client ~]# ls
anaconda-ks.cfg  compose_lnmp  constar.tar  consul  httpd  myhttpd  myh-w  mysshd1  nginx  sshd  systemctl  tomcat
[root@client ~]#
```

图 2-37　查看 Client 主机当期目录情况

(5) 将文件 myhttpd 导入为本地镜像，操作命令如下：

```
[root@client ~]# docker images
[root@client ~]# cat myhttpd |docker import - myhttpd:v1
[root@client ~]# docker images
```

命令运行结果如图 2-38 所示。

```
[root@client ~]# docker images
REPOSITORY      TAG       IMAGE ID       CREATED       SIZE
centos/httpd    latest    2cc07fbb5000   5 years ago   258MB
[root@client ~]# cat myhttpd |docker import - myhttpd:v1
sha256:fcc436e8dbcf53137bb195df67684a62782ddc8fe6f840979036891b22807f99
[root@client ~]# docker images
REPOSITORY      TAG       IMAGE ID       CREATED          SIZE
myhttpd         v1        fcc436e8dbcf   10 seconds ago   235MB
centos/httpd    latest    2cc07fbb5000   5 years ago      258MB
[root@client ~]#
```

图 2-38　将文件导入为本地镜像

(6) 启动并进入容器，运行 ls 命令，并退出容器，操作命令如下：

```
[root@client ~]# docker run -it fcc436e8dbcf /bin/bash
[root@ccca6e73fb3c /]# ls
[root@ccca6e73fb3c /]# exit
[root@client ~]# docker ps -a
```

命令运行结果如图 2-39 所示。

```
[root@client ~]# docker run -it fcc436e8dbcf /bin/bash
[root@ccca6e73fb3c /]# ls
anaconda-post.log  bin  boot  dev  etc  home  lib  lib64  media  mnt  opt  proc  root  run  run-httpd.sh  sbin  srv  sys  tmp  usr  var
[root@ccca6e73fb3c /]# exit
exit
[root@client ~]# docker ps -a
CONTAINER ID   IMAGE          COMMAND       CREATED          STATUS                     PORTS   NAMES
ccca6e73fb3c   fcc436e8dbcf   "/bin/bash"   52 seconds ago   Exited (0) 20 seconds ago          laughing_wilbur
[root@client ~]#
```

图 2-39　启动并进入容器

注意：如果虚拟机上仍有正在持续运行的容器，一定记得将其删除，否则会消耗虚拟机资源，从而影响虚拟机的性能。

双创视角

Linux 操作系统的诞生

1991 年，芬兰赫尔辛基大学计算机系的一名学生 Linus Benedict Torvalds，在学校学习操作系统课程时，受自由软件运动的精神领袖、GNU 计划以及自由软件基金会 (FSF) 的创立者 Richard Matthew Stallman 博士的影响，将自己开发的一个简陋的操作系统内核放到了互联网上。当时 GNU 计划的 Unix 操作系统正好缺少一个好用的内核，Linus Benedict Torvalds 的操作系统内核被称为 Linus's Unix，简称为 Linux，于是 Linux 诞生了。

Linux 被放到网上后，很多人陆续加入到 Linux 的开发中，他们通过网络协作推进 Linux 的发展。1994 年 3 月 Linux 1.0 发布，Linus 小时候被企鹅攻击过，因此他给 Linux 的 Logo 选择了企鹅。

Linux 继承了 Unix 的设计思想，支持多用户、多任务和多线程，具有高度的可定制性、稳定性和安全性。随着 Linux 的不断发展，其功能日渐强大，它被广泛应用到了全球各地，成为人们普遍欢迎的服务器操作系统。

项 目 小 结

本项目介绍了 Docker 容器技术、计算机虚拟化技术以及 Docker 容器与虚拟机的比较等基础知识，通过 Docker 安装、镜像和容器等基本操作，让读者掌握 Docker 容器技术的基本应用。

习 题 测 试

一、单选题

1. 云计算的服务类型通常分为 (　　)、PaaS、SaaS 三类。

A. Nginx　　B. IaaS

C. Apache　　D. Linux

2. 容器即服务 (CaaS) 是一款云服务，使用基于容器的抽象来管理和部署 (　　)。

A. 硬件　　B. 存储

C. 应用　　D. 网络

3. 由于容器不需要进行硬件虚拟以及运行完整的 (　　) 等额外开销，因此 Docker 对系统资源的利用率更高。

A. 系统软件　　B. 应用软件

C. 服务系统　　D. 操作系统

二、多选题

1. 使用容器的好处有可移植性好、(　　)。

A. 可扩展性强　　B. 高效性
C. 更高的安全性　　D. 速度快

2. 虚拟化技术需要 (　　) 等的完整支持。

A. CPU　　B. 主板芯片组
C. BIOS　　D. 软件

3. 伪终端为其他程序提供终端形式的 (　　)，但没有对应的 (　　) 设备。

A. 接口　　B. 存储
C. 软件　　D. 硬件

三、简答题

1. 简述什么是虚拟化技术。

2. 简述 Docker 容器和虚拟机有何不同。

项目三
管理 Docker 数据与网络通信

学习目标

(1) 了解 Docker 数据存储。
(2) 认识数据卷与数据卷容器。
(3) 理解 Docker 网络通信。
(4) 掌握 Docker 数据管理的方法。
(5) 掌握端口映射的方法。
(6) 掌握容器互联的方法。
(7) 掌握自定义网络的方法。

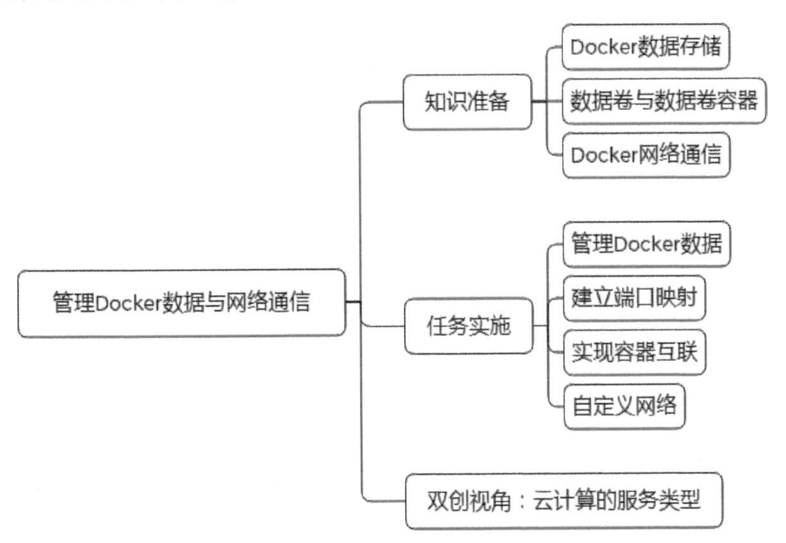

3.1 知识准备

3.1.1 Docker 数据存储

Docker 是一种使用容器来封装应用程序和其依赖关系的技术。在 Docker 中，容器是一个独立的、可执行的软件包，包含所需的操作代码、系统工具、库和设置等，所有应用程序和依赖关系都被打包在一个容器中，使它们易于部署在不同的环境中。

在 Docker 中，容器通常由镜像 (Images)、容器 (Container) 和元数据 (Metadata) 三个部分组成。Docker 镜像是一个 Linux 的文件系统，包含可以运行在 Linux 内核的程序以及相应的数据，它用于创建容器，包括应用程序及其所有依赖项。Docker 容器是由镜像创建的可运行实例，包含运行应用程序的所有内容。Docker 元数据是关于容器和镜像的注释，包括应用程序的版本、创建者和必要说明等信息。所有这些数据都需要有效存储。

Docker 默认存储目录为 /var/lib/docker，Docker 镜像、容器、日志等均存储于此，如图 3-1 所示。一般选择建立 LVM 逻辑卷这样的大容量分区来存储这些内容，确保 Docker 运行过程中存储目录的容量充足。

```
[root@docker ~]# cd /var/lib/docker
[root@docker docker]# ls
buildkit  containers  image  network  overlay2  plugins  runtimes  swarm  tmp  trust  volumes
```

图 3-1 Docker 默认的存储目录

3.1.2 数据卷与数据卷容器

在 Docker 中，用数据卷 (Data Volumes) 和数据卷容器 (Data Volume Containers) 两种方式，通过容器的数据管理操作，可以方便查看容器内产生的数据或者共享多个容器之间的数据。

1. 数据卷

数据卷是宿主机的一个目录或文件，数据卷可以存储应用程序的数据、配置文件或其他需要在容器之间传递和持久化的内容。数据卷直接将容器内数据映射到本地主机，可以实现容器数据的持久化、客户端和容器的数据交换以及容器间的数据交换。

数据卷可以供一个或多个容器使用，也可让本地与容器之间更高效地传递数据。数据卷是由 Docker daemon 挂载到容器中的一个目录，因此数据卷里面的内容不会因为容器的删除而丢失。

数据卷有以下特性：

(1) 数据卷可以在容器之间共享，使容器间的数据传递变得高效。

(2) 对容器内或本地数据卷内数据的修改会立即生效。

(3) 对数据卷的更新不会影响镜像，将应用和数据进行解耦。

(4) 数据卷会一直存在，当没有容器使用它时，可以安全地卸载。

数据卷是一个特殊的目录，可以不经过容器文件系统的常规层，直接在主机的文件系统上进行管理。数据卷可以在容器之间共享，并且可以持久存在，即使容器被删除，数据卷仍然存在。数据卷可以由 Docker 主机或其他容器创建和管理，而且可以在容器的生命周期中被挂载和卸载。

2. 数据卷容器

数据卷容器是指使用特定容器维护数据卷，在容器和主机、容器和容器之间共享数据，实现数据的备份和恢复。

数据卷容器是一个容器，专门提供数据卷给其他容器挂载。如果用户需要在多个容器之间共享一些持续更新的数据，最简单的方式是使用数据卷容器。

数据卷容器是专门用于创建和管理数据卷的容器。数据卷容器本身不运行应用程序，而是用作数据卷的持久化存储和共享点。操作中，可以首先创建一个数据卷容器并将其挂载到主机或其他容器中，其他容器可以通过挂载相同的数据卷容器来共享数据。数据卷容器提供了一个中心化的位置来管理和维护数据卷，使容器之间的数据共享更加方便和可控。

总之，Docker 中的数据卷和数据卷容器是在容器之间实现共享和持久化数据的两种不同的机制，具体的使用场景和需求决定了采用哪种机制。使用数据卷容器可以在容器之间共享数据或进行数据的持久化处理，并且能够更好地管理和控制数据卷；而数据卷只是简单地在容器中挂载一个目录并进行数据共享。

3.1.3　Docker 网络通信

当 Docker 启动时，会自动在主机上创建一个 docker0 虚拟网桥，如图 3-2 所示，可以将它看作一个软件交换机，使挂载到它的网口之间可以进行数据转发。Docker0 虚拟网桥会随着 Docker 启动而添加，随着 Docker 关闭而删除。

```
[root@docker ~]# ip a
1: lo: <LOOPBACK,UP,LOWER_UP> mtu 65536 qdisc noqueue state UNKNOWN group default qlen 1000
    link/loopback 00:00:00:00:00:00 brd 00:00:00:00:00:00
    inet 127.0.0.1/8 scope host lo
       valid_lft forever preferred_lft forever
    inet6 ::1/128 scope host
       valid_lft forever preferred_lft forever
2: ens33: <BROADCAST,MULTICAST,UP,LOWER_UP> mtu 1500 qdisc pfifo_fast state UP group default qlen 1000
    link/ether 00:0c:29:21:e8:30 brd ff:ff:ff:ff:ff:ff
    inet 192.168.1.10/24 brd 192.168.1.255 scope global ens33
       valid_lft forever preferred_lft forever
    inet6 fe80::20c:29ff:fe21:e830/64 scope link
       valid_lft forever preferred_lft forever
3: br-4e95e1fe45e9: <NO-CARRIER,BROADCAST,MULTICAST,UP> mtu 1500 qdisc noqueue state DOWN group default
    link/ether 02:42:17:71:9e:42 brd ff:ff:ff:ff:ff:ff
    inet 172.19.0.1/16 brd 172.19.255.255 scope global br-4e95e1fe45e9
       valid_lft forever preferred_lft forever
```

图 3-2　主机上的 docker0 虚拟网桥

Docker 随机分配一个本地未占用的私有网段中的一个 IP 地址给 docker0 接口，此后再启动容器时，其内的网口也会自动分配一个同网段的 IP 地址。当启动一个 Docker 容器的时候，会同时创建一对 veth pair 接口，数据包发送到一个接口时，另外一个接口也可以收到相同的数据包。这对接口一端在容器内 (即 eth0)，另一端在本地并被挂载到 docker0 网桥 (名称以 veth 开头)。这样一来，主机和容器以及容器之间都可以相互通信。Docker 在主机和所有容器之间所创建的虚拟共享网络如图 3-3 所示。

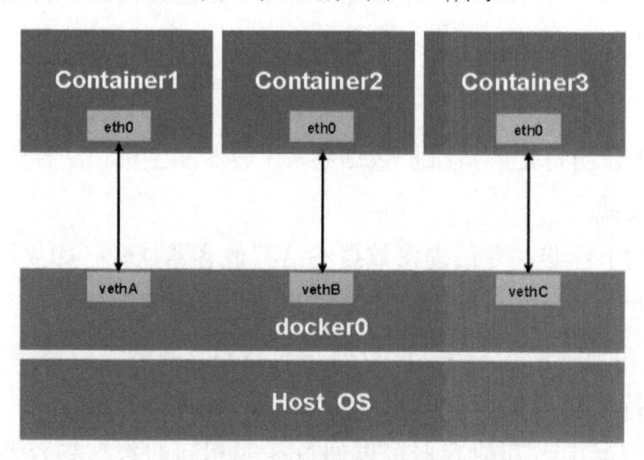

图 3-3　Docker 虚拟共享网络

在 Docker 中，容器端口可以映射到宿主机，也可以通过容器互联来为容器提供网络服务。

1. 端口映射

在启动容器时，需要指定宿主机对应的端口，否则在容器外将无法通过网络来访问容器内的服务。端口映射机制将容器内的服务提供给外部网络访问，通过将宿主机的端口映射到容器中，使外部网络访问宿主机的端口便可以获得容器内的服务。

实现端口映射，运行 docker run 命令时使用 -P(大写) 选项，随机映射一个端口，或者使用 -p(小写) 选项指定要映射的端口，将宿主机的端口映射到容器内部开放的网络端口，从而访问到容器内部应用提供的服务。

命令格式如下：

```
docker run -d -P 镜像名称，或者
docker run -d -p 指定端口：容器内服务端口 镜像名称
```

2. 容器互联

容器互联是通过容器的名称在容器间建立一条专门的网络通信隧道。简而言之，就是在源容器和接收容器之间建立一条隧道，接收容器可以看到源容器中指定的信息。

运行 docker run 命令时，使用 --link 选项可以实现容器之间的互联通信。

命令格式如下：

```
--link name:alias
```

其中，name 是要连接的容器的名称，alias 是该连接的别名。

两个容器建立了互联，即 Docker 在两个互联容器之间建立了一条安全隧道，不用映射它们的端口到宿主机上，这样避免暴露端口到外部网络，使容器内部应用的安全问题得到了一定保障。

3. 自定义网络

Docker 默认使用 docker0 作为基础网络服务，多个容器之间使用 --link 连接，使其能够通过名称互访，但是容器依赖较多且双向绑定时，此方法就比较麻烦，而使用自定义网络会更方便，Docker 自定义网络容器之间可以直接通过容器名互相访问。

命令格式如下：

docker network create [OPTIONS] 网络名

可选参数：

--subnet：设置子网范围。

--gateway：设置子网网关。

--driver：设置网络类型，可选值见表 3-1 所示。

表 3-1 Docker 自定义网络类型

网络模式	使用方法	说 明
Bridge	默认	与宿主机网络之间使用桥接模式
host	--driver host	与宿主机之间共享网络 (直连 NAT)
container	--driver container:[容器名 /ID]	与指定容器之间共享网络
none	--driver none	不设置网络

3.2 任 务 实 施

3.2.1 管理 Docker 数据

管理 Docker 数据

1. 任务目标

掌握容器的数据管理操作。

2. 任务内容

(1) 创建数据卷。

(2) 挂载主机目录为数据卷。

(3) 运用数据卷容器。

3. 完成任务所需的设备和软件

(1) 一台安装 Windows 10 操作系统的计算机。

(2) VMware Workstation、Docker。

(3) 远程管理工具 MobaXterm。

4. 任务实施步骤

第一步，创建容器 conht，并且创建两个数据卷分别挂载到 data1 和 data2 目录上，操作命令如下：

```
[root@docker ~]# docker run -d -v /data1 -v /data2 --name conht centos/httpd
[root@docker ~]# docker ps -a
[root@docker ~]# docker exec -it conht /bin/bash
[root@e8b5188fb2a6 /]# ls
```

命令运行结果如图 3-4 所示。

```
[root@docker ~]# docker run -d -v /data1 -v /data2 --name conht centos/httpd
e8b5188fb2a6f2cf1080ba5032d13613f4bb41b385fa33446dcb5400a47db36d
[root@docker ~]# docker ps -a
CONTAINER ID   IMAGE          COMMAND            CREATED         STATUS         PORTS    NAMES
e8b5188fb2a6   centos/httpd   "/run-httpd.sh"    7 seconds ago   Up 5 seconds   80/tcp   conht
[root@docker ~]# docker exec -it conht /bin/bash
[root@e8b5188fb2a6 /]# ls
anaconda-post.log  boot   data2  etc   lib    media  opt   root  run-httpd.sh  srv  tmp  var
bin                data1  dev    home  lib64  mnt    proc  run   sbin          sys  usr
```

图 3-4　创建数据卷

第二步，创建容器 conce，并将宿主机目录 /var/www 挂载到容器的 data3 目录上，即挂载主机目录为数据卷，实现宿主机与容器之间的数据迁移，操作命令如下：

```
[root@docker ~]# docker run -d -v /var/www:/data3 --name conce centos/httpd
[root@docker ~]# docker ps -a
```

命令运行结果如图 3-5 所示。

```
[root@docker ~]# docker run -d -v /var/www:/data3 --name conce centos/httpd
1a268cfedf9e6d19f9e91c5791da846b5f301305bee4eb3fc920356d5ad1b006
[root@docker ~]# docker ps -a
CONTAINER ID   IMAGE          COMMAND            CREATED          STATUS          PORTS    NAMES
1a268cfedf9e   centos/httpd   "/run-httpd.sh"    12 seconds ago   Up 10 seconds   80/tcp   conce
```

图 3-5　挂载主机目录为数据卷

第三步，在宿主机目录 /var/www 中创建一个文件 fileab，进入运行容器的挂载目录 data3 中可查看数据迁移情况，操作命令如下：

```
[root@docker ~]# cd /var/www
[root@docker www]# touch fileab
[root@docker www]# ls
[root@docker www]# cd
[root@docker ~]# docker exec -it conce /bin/bash
[root@1a268cfedf9e /]# ls
[root@1a268cfedf9e /]# cd data3
[root@1a268cfedf9e data3]# ls
```

命令运行结果如图 3-6 所示。

```
[root@docker ~]# cd /var/www
[root@docker www]# touch fileab
[root@docker www]# ls
fileab
[root@docker www]# cd
[root@docker ~]# docker exec -it conce /bin/bash
[root@1a268cfedf9e /]# ls
anaconda-post.log  boot   dev   home  lib64  mnt   proc  run           sbin  sys  usr
bin                data3  etc   lib   media  opt   root  run-httpd.sh  srv   tmp  var
[root@1a268cfedf9e /]# cd data3
[root@1a268cfedf9e data3]# ls
fileab
```

图 3-6　宿主机与容器之间的数据迁移

第四步，使用第一步创建好的容器 conht，挂载其中的数据卷到新的容器 cchtt，操作命令如下：

```
[root@docker ~]# docker run -it --volumes-from conht --name cchtt centos/httpd /bin/bash
[root@8a75e66cfc6f /]# ls
```

命令运行结果如图 3-7 所示。

```
[root@docker ~]# docker run -it --volumes-from conht --name cchtt centos/httpd /bin/bash
[root@8a75e66cfc6f /]# ls
anaconda-post.log  boot   data2  etc   lib    media  opt   root  run-httpd.sh  srv  tmp  var
bin                data1  dev    home  lib64  mnt    proc  run   sbin          sys  usr
```

图 3-7　挂载数据卷容器到新容器

第五步，在容器 cchtt 的数据卷目录 /data1 中，创建一个文件 filecd，在容器 conht 的目录 /data1 中查看情况，操作命令如下：

```
[root@8a75e66cfc6f /]# cd data1
[root@8a75e66cfc6f data1]# touch filecd
[root@8a75e66cfc6f data1]# ls
[root@8a75e66cfc6f data1]# exit
[root@docker ~]# docker exec -it conht /bin/bash
[root@e8b5188fb2a6 /]# cd data1
[root@e8b5188fb2a6 data1]# ls
```

命令运行结果如图 3-8 所示。

```
[root@8a75e66cfc6f /]# cd data1
[root@8a75e66cfc6f data1]# touch filecd
[root@8a75e66cfc6f data1]# ls
filecd
[root@8a75e66cfc6f data1]# exit
exit
[root@docker ~]# docker exec -it conht /bin/bash
[root@e8b5188fb2a6 /]# cd data1
[root@e8b5188fb2a6 data1]# ls
filecd
```

图 3-8　数据卷容器的数据迁移

3.2.2　建立端口映射

建立端口映射

1. 任务目标

掌握 Docker 网络通信的端口映射操作。

2. 任务内容

(1) 启动容器。

(2) 随机端口映射。

(3) 指定端口映射。

3. 完成任务所需的设备和软件

(1) 一台安装 Windows 10 操作系统的计算机。

(2) VMware Workstation、Docker。

(3) 远程管理工具 MobaXterm。

4. 任务实施步骤

第一步，启动容器，并随机映射一个本机端口到容器内部的服务，操作命令如下：

```
[root@docker ~]# docker run -d -P centos/httpd
[root@docker ~]# docker ps
```

命令运行结果如图 3-9 所示，可以看到，本机的 49154 端口被映射到了容器中的 80 端口。

```
[root@docker ~]# docker run -d -P centos/httpd
68f64bf3fa7682b7808061e90e11c331bd298c5191f67767420bead11a08fa71
[root@docker ~]# docker ps
CONTAINER ID   IMAGE          COMMAND            CREATED          STATUS          PORTS                                      NAMES
68f64bf3fa76   centos/httpd   "/run-httpd.sh"    13 seconds ago   Up 13 seconds   0.0.0.0:49154->80/tcp, :::49154->80/tcp   adoring_ch
aplygin
[root@docker ~]#
```

图 3-9　随机映射一个本机端口到容器内部的服务

第二步，在浏览器地址栏输入 http://192.168.1.10:49154，即访问宿主机的 49154 端口，就可以访问到容器内部 web 服务提供的界面，如图 3-10 所示。

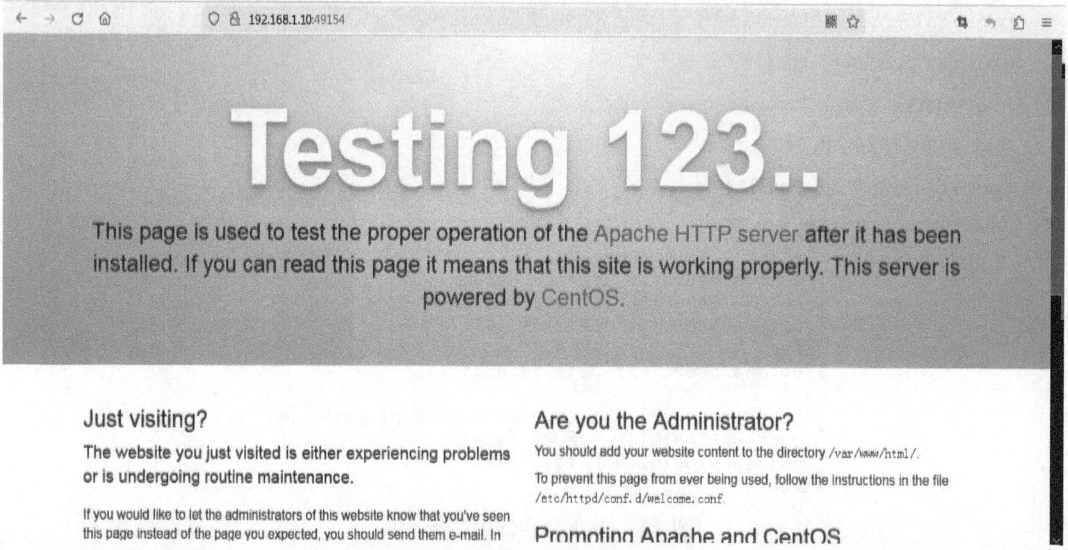

192.168.1.10:49154

Testing 123..

This page is used to test the proper operation of the Apache HTTP server after it has been installed. If you can read this page it means that this site is working properly. This server is powered by CentOS.

Just visiting?

The website you just visited is either experiencing problems or is undergoing routine maintenance.

If you would like to let the administrators of this website know that you've seen this page instead of the page you expected, you should send them e-mail. In

Are you the Administrator?

You should add your website content to the directory /var/www/html/.

To prevent this page from ever being used, follow the instructions in the file /etc/httpd/conf.d/welcome.conf.

Promoting Apache and CentOS

图 3-10　访问宿主机的 49154 端口

第三步，指定一个本机端口映射到容器内部的服务，操作命令如下：

```
[root@docker ~]# docker run -d -p 49500:80 centos/httpd
[root@docker ~]# docker ps
```

命令运行结果如图 3-11 所示，可以看到，本机的 49500 端口被映射到了容器中的 80 端口。

```
[root@docker ~]# docker run -d -p 49500:80  centos/httpd
02eabbfc300bd8e6f7394462eddc0a19184da64c293bbd464f5e2fabe11bf533
[root@docker ~]# docker ps
CONTAINER ID   IMAGE          COMMAND            CREATED          STATUS          PORTS                                        NAMES
02eabbfc300b   centos/httpd   "/run-httpd.sh"    9 seconds ago    Up 7 seconds    0.0.0.0:49500->80/tcp, :::49500->80/tcp    practical_
williams
68f64bf3fa76   centos/httpd   "/run-httpd.sh"    32 minutes ago   Up 32 minutes   0.0.0.0:49154->80/tcp, :::49154->80/tcp    adoring_ch
aplygin
[root@docker ~]#
```

图 3-11　指定一个本机端口映射到容器内部的服务

第四步，在浏览器地址栏输入 http://192.168.1.10:49500，即访问宿主机的 49500 端口，就可以访问到容器内部 web 服务提供的界面，如图 3-12 所示。

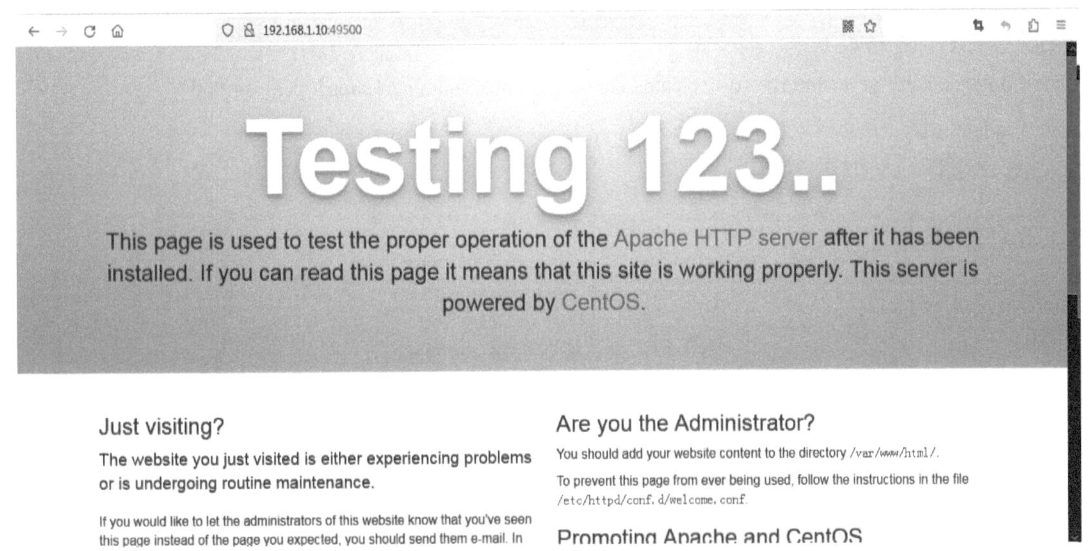

图 3-12　访问宿主机的 49500 端口

3.2.3　实现容器互联

实现容器互联

1. 任务目标

掌握 Docker 容器互联的操作。

2. 任务内容

(1) 创建源容器。

(2) 创建接收容器。

(3) 测试容器互联。

3. 完成任务所需的设备和软件

(1) 一台安装 Windows 10 操作系统的计算机。

(2) VMware Workstation、Docker。

(3) 远程管理工具 MobaXterm。

4. 任务实施步骤

第一步，创建名称为 container1 的源容器，操作命令如下：

```
[root@docker ~]# docker run -d -P --name container1 centos/httpd
[root@docker ~]# docker ps
```

命令运行结果如图 3-13 所示。

```
[root@docker ~]# docker run -d -P --name container1 centos/httpd
0024e2b3dcbe9f8e6397321680f7470b101a2a20121758d895b8bc2945d25690
[root@docker ~]# docker ps
CONTAINER ID   IMAGE          COMMAND            CREATED          STATUS          PORTS                                     NAMES
0024e2b3dcbe   centos/httpd   "/run-httpd.sh"    56 seconds ago   Up 55 seconds   0.0.0.0:49155->80/tcp, :::49155->80/tcp   container1
[root@docker ~]#
```

图 3-13　创建源容器 container1

第二步，创建名称为 container2 的接收容器，使用 --link 选项指定连接容器，实现容器互联，操作命令如下：

```
[root@docker ~]# docker run -d -P --name container2 -link container1:web1  centos/httpd
[root@docker ~]# docker ps
```

命令运行结果如图 3-14 所示。

```
[root@docker ~]# docker run -d -P --name container2 --link container1:web1  centos/httpd
7d1763eeb94c58d5b0a6d6bdb9aecaa84264bf4c446a0dcf909b99ca73dcf5d0
[root@docker ~]# docker ps
CONTAINER ID   IMAGE          COMMAND            CREATED         STATUS         PORTS                                     NAMES
7d1763eeb94c   centos/httpd   "/run-httpd.sh"    9 seconds ago   Up 7 seconds   0.0.0.0:49156->80/tcp, :::49156->80/tcp   container2
0024e2b3dcbe   centos/httpd   "/run-httpd.sh"    8 minutes ago   Up 8 minutes   0.0.0.0:49155->80/tcp, :::49155->80/tcp   container1
[root@docker ~]# ^C
```

图 3-14　创建接收容器 container2

第三步，测试容器互联，进入容器查看容器之间连通性，操作命令如下：

```
[root@docker ~]# docker exec -it container2 /bin/bash
[root@7d1763eeb94c /]# ping container1
```

命令运行结果如图 3-15 所示，可以看到，容器 container2 和 container1 已经建立了互联。

```
[root@docker ~]# docker exec -it container2 /bin/bash
[root@7d1763eeb94c /]# ping container1
PING web1 (172.17.0.2) 56(84) bytes of data.
64 bytes from web1 (172.17.0.2): icmp_seq=1 ttl=64 time=0.094 ms
64 bytes from web1 (172.17.0.2): icmp_seq=2 ttl=64 time=0.068 ms
64 bytes from web1 (172.17.0.2): icmp_seq=3 ttl=64 time=0.071 ms
64 bytes from web1 (172.17.0.2): icmp_seq=4 ttl=64 time=0.072 ms
64 bytes from web1 (172.17.0.2): icmp_seq=5 ttl=64 time=0.099 ms
64 bytes from web1 (172.17.0.2): icmp_seq=6 ttl=64 time=0.074 ms
^C
--- web1 ping statistics ---
6 packets transmitted, 6 received, 0% packet loss, time 4999ms
rtt min/avg/max/mdev = 0.068/0.079/0.099/0.015 ms
[root@7d1763eeb94c /]#
```

图 3-15　测试容器互联

3.2.4　自定义网络

自定义网络

1. 任务目标

掌握 Docker 自定义网络的操作。

2. 任务内容

(1) 创建一个网络。

(2) 创建两个容器，并连接到所创建的网络上。

(3) 测试容器的连通性。

3. 完成任务所需的设备和软件

(1) 一台安装 Windows 10 操作系统的计算机。

(2) VMware Workstation、Docker。

(3) 远程管理工具 MobaXterm。

4. 任务实施步骤

第一步，创建一个 mynet 网络，操作命令如下：

```
[root@docker ~]# docker network create mynet
```

命令运行结果如图 3-16 所示。

```
[root@docker ~]# docker network create mynet
0b0d972bfb565e87fca56081fbe1fa8c36a88db2e1d580f7e0e62481130adceb
```

图 3-16　创建一个 mynet 网络

第二步，启动两个容器，并连接到 mynet 网络，操作命令如下：

```
[root@docker ~]# docker run -d -P --name cweb1 --network mynet centos/httpd
[root@docker ~]# docker run -d -P --name cweb2 --network mynet centos/httpd
[root@docker ~]# docker ps
```

命令运行结果如图 3-17 所示。

```
[root@docker ~]# docker run -d -P --name cweb1 --network mynet centos/httpd
0ee9ea639daf859f2b01fcd178ea95e830dab037f84fd11c324be665c84faa19
[root@docker ~]# docker run -d -P --name cweb2 --network mynet centos/httpd
b8ec07c4eed095b0a52b576f3f89903cce2f992932a7d3b811e542d67bdd7e6d
[root@docker ~]# docker ps
CONTAINER ID   IMAGE          COMMAND            CREATED          STATUS          PORTS                                       NAMES
b8ec07c4eed0   centos/httpd   "/run-httpd.sh"    14 seconds ago   Up 12 seconds   0.0.0.0:49160->80/tcp, :::49160->80/tcp     cweb2
0ee9ea639daf   centos/httpd   "/run-httpd.sh"    30 seconds ago   Up 28 seconds   0.0.0.0:49159->80/tcp, :::49159->80/tcp     cweb1
[root@docker ~]#
```

图 3-17　启动两个容器并连接到 mynet 网络

第三步，进入容器 cweb1，测试与容器 cweb2 的连通性，操作命令如下：

```
[root@docker ~]# docker exrc -it cweb1 /bin/bash
[root@0ee9ea639daf /]# ping cweb2
```

命令运行结果如图 3-18 所示，可以看出两个容器是联通的。

```
[root@docker ~]# docker exec -it cweb1 /bin/bash
[root@0ee9ea639daf /]# ping cweb2
PING cweb2 (172.21.0.3) 56(84) bytes of data.
64 bytes from cweb2.mynet (172.21.0.3): icmp_seq=1 ttl=64 time=0.063 ms
64 bytes from cweb2.mynet (172.21.0.3): icmp_seq=2 ttl=64 time=0.051 ms
64 bytes from cweb2.mynet (172.21.0.3): icmp_seq=3 ttl=64 time=0.056 ms
64 bytes from cweb2.mynet (172.21.0.3): icmp_seq=4 ttl=64 time=0.058 ms
64 bytes from cweb2.mynet (172.21.0.3): icmp_seq=5 ttl=64 time=0.048 ms
^C
--- cweb2 ping statistics ---
5 packets transmitted, 5 received, 0% packet loss, time 4001ms
rtt min/avg/max/mdev = 0.048/0.055/0.063/0.007 ms
[root@0ee9ea639daf /]# ^C
```

图 3-18　测试容器 cweb1 与容器 cweb2 的连通性

第四步，进入容器 cweb2，测试与容器 cweb1 的连通性，操作命令如下：

```
[root@docker ~]# docker exrc -it cweb2 /bin/bash
[root@b8ec07c4eed0 /]# ping cweb1
```

命令运行结果如图 3-19 所示，可以看出两个容器是联通的。

```
[root@docker ~]# docker exec -it cweb2 /bin/bash
[root@b8ec07c4eed0 /]# ping cweb1
PING cweb1 (172.21.0.2) 56(84) bytes of data.
64 bytes from cweb1.mynet (172.21.0.2): icmp_seq=1 ttl=64 time=0.050 ms
64 bytes from cweb1.mynet (172.21.0.2): icmp_seq=2 ttl=64 time=0.051 ms
64 bytes from cweb1.mynet (172.21.0.2): icmp_seq=3 ttl=64 time=0.051 ms
64 bytes from cweb1.mynet (172.21.0.2): icmp_seq=4 ttl=64 time=0.067 ms
64 bytes from cweb1.mynet (172.21.0.2): icmp_seq=5 ttl=64 time=0.051 ms
64 bytes from cweb1.mynet (172.21.0.2): icmp_seq=6 ttl=64 time=0.051 ms
^C
--- cweb1 ping statistics ---
6 packets transmitted, 6 received, 0% packet loss, time 5000ms
rtt min/avg/max/mdev = 0.050/0.053/0.067/0.009 ms
[root@b8ec07c4eed0 /]#
```

图 3-19　测试容器 cweb2 与容器 cweb1 的连通性

双创视角

云计算的服务类型

云计算是信息时代的一大飞跃，是继计算机、互联网之后的又一革新。云指的是互联网，云计算是一种通过互联网以服务的方式提供动态可伸缩的虚拟化资源的计算模式。

云计算通常提供三种类型的服务，它们是基础设施即服务 (Infrastructure as a Service，IaaS)、平台即服务 (Platform as a Service，PaaS) 和软件即服务 (Software as a Service，SaaS)。

基础设施即服务 (IaaS)：最基本的云计算服务类型，为用户提供计算、存储和网络等基础设施资源。用户可以通过云平台租赁或购买这些资源，从而构建和管理自己的虚拟数据中心。用户可以根据需求配置和使用计算资源，成本相对较低，但用户需要自行管理和维护相应的基础设施资源，因此对技术能力的要求较高。

平台即服务 (PaaS)：为用户提供应用程序开发和部署所需的平台和工具，用户可以通过云平台租赁或购买这些服务和工具，然后进行应用程序的开发、测试和部署。用户无需购买和维护软件开发所需的硬件和软件资源，因此降低了开发和维护成本。

软件即服务 (SaaS)：为用户提供应用程序服务，用户可以通过云平台使用这些应用程序。用户无需购买和维护应用程序所需的硬件和软件资源，因此降低了使用和维护成本。

项目小结

本项目介绍了 Docker 的数据存储、数据卷与数据卷容器、Docker 网络通信等知识，完成了 Docker 的数据管理、端口映射、容器互联、自定义网络等操作任务，为进一步掌握 Docker 容器技术做好铺垫。

习题测试

一、单选题

1. Docker 是一种使用 (　　) 来封装应用程序及其依赖关系的技术。

A. 硬件　　　　B. 软件

C. 数据　　　　D. 容器

2. 通过容器的（　　）操作，可以方便查看容器内产生的数据或者共享多个容器之间的数据。

A. 导入　　B. 导出

C. 数据管理　　D. 数据通信

3. 当 Docker 启动时，会自动在主机上创建一个 docker0 虚拟网桥，可以将它看作一个软件（　　），使挂载到它的网口之间进行数据转发。

A. 中继器　　B. 集线器

C. 交换机　　D. 路由器

二、多选题

1. Docker 默认存储目录 /var/lib/docker 中，一般存储（　　）等信息。

A. Docker 镜像　　B. Docker 容器

C. Docker 日志　　D. Docker 仓库

2. 数据卷可以存储应用程序的数据、配置文件或其他需要在容器之间（　　）和（　　）的内容。

A. 存储　　B. 虚拟化

C. 传递　　D. 持久化

3. Docker 默认使用 docker0 作为基础网络服务，多个容器之间使用（　　）连接，使其能够通过（　　）互访。

A. --create　　B. --link

C. 名称　　D. 标签

三、简答题

1. 数据卷和数据卷容器一样吗？阐明理由。

2. 简述在 Docker 中容器如何提供网络服务。

项目四

创建 Docker 镜像

学习目标

(1) 认识 Docker 镜像的结构。
(2) 了解创建 Docker 镜像的方法。
(3) 认识 Dockerfile。
(4) 掌握通过容器创建镜像的方法。
(5) 掌握通过 Dockerfile 构建镜像的方法。

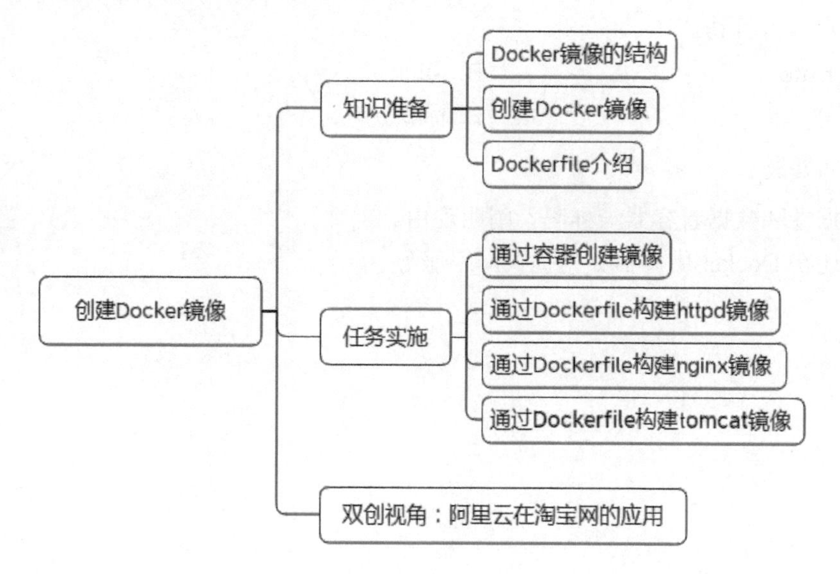

4.1 知识准备

4.1.1 Docker 镜像的结构

镜像是一个轻量级、可执行的独立软件包，用来打包软件运行环境和基于运行环境开发的软件。镜像包含运行某个软件所需的所有内容，比如代码、运行所需的库、环境变量和配置文件等。

Docker 镜像先将镜像仓库中的镜像下载到本地 Docker 主机上，然后启动本地某个镜像来得到一个或多个容器，进入容器之后，就可以在其中进行相关操作。如果要迁移容器中已经安装好的服务，就需要将环境及搭建的应用服务生成新的镜像。镜像由多个层组成，每层叠加之后将形成一个独立的对象。在拉取镜像的过程中可以看到镜像的多个层，如图 4-1 所示。

```
[root@docker ~]# docker pull daocloud.io/library/logstash
Using default tag: latest
latest: Pulling from library/logstash
bc9ab73e5b14: Pull complete
193a6306c92a: Pull complete
e5c3f8c317dc: Pull complete
d21441932c53: Pull complete
fa76b0d25092: Pull complete
346fd8610875: Pull complete
3ca5d6af9022: Pull complete
c06cfa2cea32: Pull complete
c3b4c0fc322d: Pull complete
03439cff2d79: Pull complete
1b61d3b5e376: Pull complete
e064ae74a9ef: Pull complete
b9894fbeaa66: Pull complete
b284b8c3b8f0: Pull complete
81e6e144df62: Pull complete
ad1e245c0b9b: Pull complete
Digest: sha256:ca002d3616a134a775da04469a1c7292299d0260cfec825ffb646df4a02d8f77
Status: Downloaded newer image for daocloud.io/library/logstash:latest
daocloud.io/library/logstash:latest
```

图 4-1　拉取镜像的过程

镜像是一个只读文件，容器在镜像各层的最上面加了一个读写层，在容器里进行的所有文件改动都会记录到该读写层中，并不会修改镜像的各层。如果删除容器，就删除了这个读写层，文件改动也随之丢失。镜像的分层结构如图 4-2 所示。

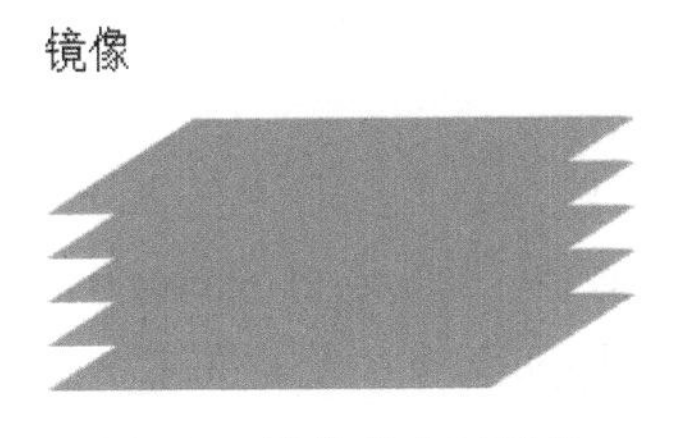

图 4-2　镜像的分层结构

Docker 已经成为现代应用程序开发和部署的必备工具之一，Docker 镜像将应用程序及其依赖项封装在一个容器中，使它们能够在任何地方运行。Docker 镜像的分层结构具有以下优点：

(1) Docker 在每个镜像层中只存储差异，每个新的镜像层都可以利用之前的层，这样减少了构建过程中的重复工作，从而能够加快构建过程，提高开发效率。

(2) Docker 镜像的大小通常比传统的虚拟机镜像小很多，使得 Docker 容器能够更快速地部署和启动，也更易于迁移和存储。

(3) 每个 Docker 镜像层都是单独管理的，如果需要更新应用程序的某个数据，只需在该数据所在的镜像层中进行更改即可，不用重新构建整个镜像，所以 Docker 镜像容易扩展和维护。

4.1.2 Docker 镜像的创建方法

当运行容器时，使用的镜像如果不在本地，Docker 就会自动从 Docker 镜像仓库服务中下载。一般默认从 Docker Hub 公共镜像源下载，当然也可以从其他镜像仓库服务中下载，比如国内的镜像仓库等。镜像除了可以在线下载，也可以在本地制作镜像，比如，需要用到符合特定需求的专用镜像时，可以通过某种特定方法创建此镜像。

1. 基于容器创建镜像

将一个容器中运行的程序及其运行环境打包生成新的镜像，即在运行的容器中做一些修改操作，然后把这些操作保存到生成的镜像中。

2. 基于模板创建镜像

首先通过 OpenVZ 开源项目下载操作系统的模板文件，然后导入该模板文件就可以生成镜像。

OpenVZ 是一种开源虚拟化技术，可以在单台物理服务器上运行多个虚拟化操作系统，并使用容器技术进行隔离，这些隔离的虚拟化操作系统称为虚拟专用服务器 (Virtual Private Server，VPS)。每个 VPS 的运行和独立服务器完全一致，它拥有自己的用户、IP 地址、内存、处理器、系统库文件和配置文件等，但是共享操作系统内核。由于 OpenVZ 的轻量级处理损耗和高效设计，常用于运行应用服务和实时数据生产型服务器虚拟化。

3. 基于 Dockerfile 创建镜像

Docker 可以通过读取 Dockerfile 中的指令自动创建镜像。Dockerfile 是一个文本文件，其中包含了一条条创建镜像所需的指令和说明，每条指令都会创建一个新的镜像层，每一条指令的内容就是描述该镜像层应当如何创建的。需要注意的是，Dockerfile 有其特定的语法规则，在编写脚本时一定要遵守。

4.1.3 Dockerfile 介绍

Dockerfile 创建 Docker 镜像的过程，就是把 Linux 操作命令写到了 Dockerfile 脚本中，通过 Docker build 去执行设置好的操作命令，最终创建出新的镜像。

Dockerfile 文件由四部分构成，分别是基础镜像信息、维护者信息、镜像操作指令和容器启动时的执行命令。Dockerfile 的常用操作指令见表 4-1 所示。

表 4-1 Dockerfile 操作指令

序号	指 令	含 义
1	FROM	指定基础镜像，并且必须是第一条指令。如果不以任何镜像为基础，则指令为 FROM scratch
2	MAINTAINER	说明维护人的信息
3	ADD	复制源文件到镜像中
4	COPY	复制本地文件到镜像中
5	RUN	运行指定的命令
6	EXPOSE	暴露容器运行时的监听端口给外部
7	ENV	设置环境变量
8	CMD	容器启动时默认执行的命令或参数
9	ENTRYPOINT	容器启动时运行的启动命令
10	VOLUME	实现挂载，可以将宿主机目录挂载到容器中，完成持久化存储数据
11	USER	设置启动容器的用户
12	WORKDIR	设置工作目录
13	LABEL	以键值对的形式给镜像添加一些元数据
14	ARG	设置环境变量，只有在 Dockerfile 内有效
15	ONBUILD	延迟创建命令的执行

编写 Dockerfile 脚本时，第一行必须使用 FROM 指令指定基础镜像，然后使用 MAINTAINER 指令说明镜像维护者的信息，再使用 RUN、ADD、COPY 等指令完成相关操作，最后使用 CMD 等指令指定启动容器时要运行的命令。

Dockerfile 脚本中可能会用到其他文件，而这些文件要和 Dockerfile 文件在同一级父目录下，否则在生成镜像的过程中会出现找不到所需文件的情况。另外，因为 Dockerfile 的每条指令都会创建一个新的镜像层，所以尽量少写指令行，目的是简化镜像的生成过程。

4.2 任务实施

4.2.1 通过容器创建镜像

通过容器创建镜像

1. 任务目标

掌握通过容器创建本地镜像的方法。

2. 任务内容

将一个正在运行的容器直接提交为一个镜像。

3. 完成任务所需的设备和软件

(1) 一台安装 Windows 10 操作系统的计算机。

(2) VMware Workstation、Docker。

(3) 远程管理工具 MobaXterm。

4. 任务实施步骤

第一步，启动容器，并查看容器运行状态，操作命令如下：

```
[root@docker ~]# docker run -itd centos/httpd
[root@docker ~]# docker ps -a
```

命令运行结果如图 4-3 所示。

```
[root@docker ~]# docker run -itd centos/httpd
524647ea920e6fddffd416d53641a0d1c3a41fbdcc0f5277a4942a57f13263a1
[root@docker ~]# docker ps -a
CONTAINER ID   IMAGE          COMMAND              CREATED          STATUS                   PORTS     NAMES
524647ea920e   centos/httpd   "/run-httpd.sh"      5 seconds ago    Up 4 seconds             80/tcp    cool_lederberg
87e982527601   sshd:new       "/bin/bash -c ls"    4 hours ago      Exited (0) 4 hours ago             funny_brattain
```

图 4-3　启动容器并查看运行状态

第二步，进入容器，新建文件 a.txt，查看结果，最后退出容器，操作命令如下：

```
[root@docker ~]# docker exec -it 524647ea920e /bin/bash
[root@524647ea920e /]# touch a.txt
[root@524647ea920e /]# ls
[root@524647ea920e /]# exit
```

命令运行结果如图 4-4 所示。

```
[root@docker ~]# docker exec -it 524647ea920e /bin/bash
[root@524647ea920e /]# touch a.txt
[root@524647ea920e /]# ls
a.txt              bin   dev  home  lib64  mnt  proc  run            sbin  sys  usr
anaconda-post.log  boot  etc  lib   media  opt  root  run-httpd.sh   srv   tmp  var
[root@524647ea920e /]# exit
exit
[root@docker ~]# ^C
[root@docker ~]#
```

图 4-4　修改容器内容

第三步，基于第二步修改的容器，创建新镜像 myima:test 并查看结果，操作命令如下：

```
[root@docker ~]# docker commit -m "new" -a "jx" 524647ea920e myima:test
[root@docker ~]# docker images
```

命令运行结果如图 4-5 所示。

```
[root@docker ~]# docker commit -m "new" -a "jx" 524647ea920e myima:test
sha256:7926f550e82196223e8d9b7e7a6b0933697313cc981048118854eb830af9cbd7
[root@docker ~]# docker images
REPOSITORY          TAG      IMAGE ID       CREATED          SIZE
myima               test     7926f550e821   11 seconds ago   258MB
centos              mysql    7df9170a1a89   3 months ago     476MB
tomcat              centos   40552ff22f8f   3 months ago     626MB
nginx               new      b12a1a921e1f   3 months ago     471MB
local/c7-systemd    latest   12ee522a64b2   3 months ago     365MB
sshd                new      a7742ca6b111   3 months ago     365MB
```

图 4-5　创建新的镜像

第四步，利用新镜像启动容器，查看之前容器的改动是否还在，操作命令如下：

```
[root@docker ~]# docker run -itd 7926f550e821
[root@docker ~]# docker exec -it af237829 /bin/bash
[root@af237829879e /]# ls
```

命令运行结果如图 4-6 所示，可见之前容器的改动是存在的。

```
[root@docker ~]# docker run -itd 7926f550e821
af237829879e3d40d58c65f2ae0b96e8dfc2f3339a9a45b7c4a7aeebda5ec33a
[root@docker ~]# docker exec -it af237829 /bin/bash
[root@af237829879e /]# ls
a.txt              bin   dev  home  lib64  mnt  proc  run           sbin  sys  usr
anaconda-post.log  boot  etc  lib   media  opt  root  run-httpd.sh  srv   tmp  var
```

图 4-6　验证新镜像

4.2.2　通过 Dockerfile 构建 httpd 镜像

通过 Dockerfile 构建 httpd 镜像

1. 任务目标

理解通过 Dockerfile 构建 httpd 镜像的方法。

2. 任务内容

(1) 创建工作目录。

(2) 编写 Dockerfile 文件。

(3) 生成镜像。

(4) 启动容器测试。

3. 完成任务所需的设备和软件

(1) 一台安装 Windows 10 操作系统的计算机。

(2) VMware Workstation、Docker。

(3) 远程管理工具 MobaXterm。

4. 任务实施步骤

第一步，建立工作目录并进入其中，操作命令如下：

```
[root@docker ~]# mkdir httpd
[root@docker ~]# cd httpd
```

第二步，创建并编辑 Dockerfile 文件，操作命令如下：

```
[root@docker httpd]# vim Dockerfile
```

Dockerfile 文件的内容如下，数字表示行号：

```
1 FROM centos:7
2 MAINTAINER docker
3 RUN yum install httpd httpd-devel -y \
4 && echo "httpd main page" > /var/www/html/index.html
5 VOLUME ["/var/www/html/"]
6 EXPOSE  80
7 CMD ["/usr/sbin/httpd","-D","FOREGROUND"]
```

第 1 行指明基础镜像，第 2 行说明镜像维护者的信息，第 3 行安装 Apache，第 4 行在 index.html 文件中添加输出指令，第 5 行在容器中创建一个挂载点，映射到自动生成的主机目录上，第 6 行指定容器在运行时监听的端口，第 7 行指定容器启动时所要执行的命令。

“\” 表示换行。为了增强代码的可读性，可对较长代码做换行处理。

“&&”表示其前后代码一起执行。当多条代码需同时执行时，可用该符号连接多条代码。

第三步，构建镜像，操作命令如下：

```
[root@docker httpd]# docker build -t httpd:new .
```

注意：该命令最后的点表示在当前目录下，一定不要忘记。

命令运行结果如图 4-7 所示。

```
[root@docker httpd]# docker build -t httpd:new .
Sending build context to Docker daemon  2.048kB
Step 1/6 : FROM centos:7
 ---> eeb6ee3f44bd
Step 2/6 : MAINTAINER docker
 ---> Running in 4be643fbb861
Removing intermediate container 4be643fbb861
 ---> dfa2e3524b13
Step 3/6 : RUN yum install httpd httpd-devel -y && echo "httpd main page" > /var/www/html/index.html
 ---> Running in f28b41a6d780
Loaded plugins: fastestmirror, ovl
Determining fastest mirrors
 * base: mirrors.163.com
 * extras: mirrors.163.com
 * updates: mirrors.163.com
Resolving Dependencies
--> Running transaction check
---> Package httpd.x86_64 0:2.4.6-99.el7.centos.1 will be installed
--> Processing Dependency: httpd-tools = 2.4.6-99.el7.centos.1 for package: httpd-2.4.6-99.el7.centos.1.x86_64
--> Processing Dependency: system-logos >= 7.92.1-1 for package: httpd-2.4.6-99.el7.centos.1.x86_64
--> Processing Dependency: /etc/mime.types for package: httpd-2.4.6-99.el7.centos.1.x86_64
--> Processing Dependency: libaprutil-1.so.0()(64bit) for package: httpd-2.4.6-99.el7.centos.1.x86_64
Dependency Updated:
  cyrus-sasl-lib.x86_64 0:2.1.26-24.el7_9     expat.x86_64 0:2.1.0-15.el7_9
  openldap.x86_64 0:2.4.44-25.el7_9

Complete!
Removing intermediate container f28b41a6d780
 ---> 30d85cc6a551
Step 4/6 : VOLUME ["/var/www/html/"]
 ---> Running in 7fffc1c34e50
Removing intermediate container 7fffc1c34e50
 ---> 910afa23a443
Step 5/6 : EXPOSE  80
 ---> Running in befdfc6421b8
Removing intermediate container befdfc6421b8
 ---> a703f52e38a7
Step 6/6 : CMD ["/usr/sbin/httpd","-D","FOREGROUND"]
 ---> Running in 1ad17ff5ba7b
Removing intermediate container 1ad17ff5ba7b
 ---> 542540f88e3f
Successfully built 542540f88e3f
Successfully tagged httpd:new
[root@docker httpd]#
```

图 4-7　构建 httpd 镜像

第四步，查看本地镜像，操作命令如下：

```
[root@docker]# docke images
```

命令运行结果如图 4-8 所示。

```
[root@docker httpd]# docker images
REPOSITORY       TAG          IMAGE ID        CREATED          SIZE
httpd            new          542540f88e3f    19 minutes ago   505MB
tomcat           new          fd0b1833593e    5 hours ago      615MB
nginx            new          b12a1a921e1f    18 months ago    471MB
sshd             new          a7742ca6b111    18 months ago    365MB
mysql            5.7          f26e21ddd20d    18 months ago    450MB
hello-world      latest       feb5d9fea6a5    2 years ago      13.3kB
centos           7            eeb6ee3f44bd    2 years ago      204MB
centos           7.6.1810     f1cb7c7d58b7    4 years ago      202MB
centos/httpd     latest       2cc07fbb5000    4 years ago      258MB
[root@docker httpd]#
```

图 4-8　查看本地镜像

第五步，启动容器并查看，操作命令如下：

```
[root@docker ~]# docker run -d -p 8001:80 --name httpd httpd:new
[root@docker ~]# docker ps -a
```

命令运行结果如图 4-9 所示。

```
[root@docker httpd]# docker run -d -p 8001:80 --name httpd httpd:new
3707a2731ee59bb9b156dcd4ea3b2f4075bca29dc703bbd7f392d798075430e6
[root@docker httpd]# docker ps -a
CONTAINER ID   IMAGE       COMMAND                  CREATED          STATUS          PORTS                                   NAMES
3707a2731ee5   httpd:new   "/usr/sbin/httpd -D …"   22 seconds ago   Up 21 seconds   0.0.0.0:8001->80/tcp, :::8001->80/tcp   httpd
[root@docker httpd]#
```

图 4-9　启动容器

第六步，通过浏览器访问地址 192.168.1.10:8001，结果如图 4-10 所示，可见构建的 httpd 镜像测试成功。

图 4-10　访问页面

4.2.3　通过 Dockerfile 构建 nginx 镜像

通过 Dockerfile 构建 nginx 镜像

1. 任务目标

理解通过 Dockerfile 构建 nginx 镜像的方法。

2. 任务内容

(1) 创建工作目录。

(2) 编写 Dockerfile 文件。

(3) 生成镜像。

(4) 启动容器测试。

3. 完成任务所需的设备和软件

(1) 一台安装 Windows 10 操作系统的计算机。

(2) VMware Workstation、Docker。

(3) 远程管理工具 MobaXterm。

4. 任务实施步骤

第一步，建立工作目录，操作命令如下：

```
[root@docker ~]# mkdir nginx
```

```
[root@docker ~]# cd nginx
```

第二步，创建并编辑 Dockerfile 文件，操作命令如下：

```
[root@docker nginx]# vim Dockerfile
```

Dockerfile 文件的内容如下所示，共 14 行代码：

```
1  FROM centos:7
2  MAINTAINER cloud-ops@centos.org
3  RUN yum install -y wget proc-devel net-tools gcc zlib zlib-devel make openssl-devel
4  RUN wget http://nginx.org/download/nginx-1.9.7.tar.gz
5  RUN tar zxf nginx-1.9.7.tar.gz
6  WORKDIR nginx-1.9.7
7  RUN ./configure --prefix=/usr/local/nginx && make && make install
8  EXPOSE 80
9  EXPOSE 443
10  RUN echo "daemon off;">>/usr/local/nginx/conf/nginx.conf
11  WORKDIR /root/nginx
12  ADD run.sh /run.sh
13  RUN chmod 775  /run.sh
14  CMD ["/run.sh"]
```

第 1 行指明了基础镜像，第 2 行说明镜像维护者的信息，第 3 行安装相关依赖包，第 4～5 行下载并解压 nginx 源码包，第 6～7 行编译并安装 nginx，第 8～9 行开启 80 端口和 443 端口，第 10 行修改 nginx 配置文件以 daemon 方式启动，第 11～13 行复制启动脚本并设置权限，第 14 行启动容器时执行脚本。

第三步，编写执行脚本内容，操作命令如下：

```
[root@docker nginx]# vim run.sh
1  #!/bin/bash
2  /usr/local/nginx/sbin/nginx
```

第四步，构建镜像，操作命令如下：

```
[root@docker nginx]# docker build -t nginx:new .
```

第五步，启动容器并查看，操作命令如下：

```
[root@docker ~]# docker run -d -P nginx:new
```

“-P”表示 Docker 会随机映射一个端口到容器内部开放的网络端口。

```
[root@docker ~]# docker ps -a
```

命令运行结果如图 4-11 所示，可以看到随机映射的端口是 49160。

第六步，通过浏览器访问地址 192.168.1.10:49160，如图 4-12 所示，可见构建的 nginx 镜像测试成功。

```
[root@docker ~]# docker images
REPOSITORY                   TAG        IMAGE ID       CREATED         SIZE
myima                        test       7926f550e821   3 hours ago     258MB
centos                       mysql      7df9170a1a89   3 months ago    476MB
tomcat                       centos     40552ff22f8f   3 months ago    626MB
nginx                        new        b12a1a921e1f   3 months ago    471MB
local/c7-systemd             latest     12ee522a64b2   3 months ago    365MB
sshd                         new        a7742ca6b111   3 months ago    365MB
my_tomcat                    v1         70dafa8f2f2a   4 months ago    612MB
mysql                        5.7        f26e21ddd20d   4 months ago    450MB
httpd                        latest     118b6abfbf55   4 months ago    144MB
mysql                        5.6        dd3b2a5dcb48   7 months ago    303MB
hello-world                  latest     feb5d9fea6a5   10 months ago   13.3kB
centos                       7          eeb6ee3f44bd   10 months ago   204MB
centos                       7.6.1810   f1cb7c7d58b7   3 years ago     202MB
centos/httpd                 latest     2cc07fbb5000   3 years ago     258MB
guyton/centos6               latest     5f3e1df89d22   3 years ago     228MB
daocloud.io/library/logstash latest     33c2b80b5322   3 years ago     653MB
[root@docker ~]# docker run -d -P nginx:new
5e6aa708991611f7288ec203782887f8cd00f3a1da4dd61ffdf4e00bf7c50cf6
[root@docker ~]# docker ps -a
CONTAINER ID   IMAGE       COMMAND      CREATED         STATUS         PORTS
                           NAMES
5e6aa7089916   nginx:new   "/run.sh"    8 seconds ago   Up 4 seconds   0.0.0.0:49160->80/tcp, :::49160->80/tcp, 0.0.0.0:49159->
443/tcp, :::49159->443/tcp   xenodochial_chebyshev
```

图 4-11　启动容器

图 4-12　访问页面

4.2.4　通过 Dockerfile 构建 tomcat 镜像

通过 Dockerfile 构建 tomcat 镜像

1. 任务目标

理解通过 Dockerfile 构建 tomcat 镜像的方法。

2. 任务内容

(1) 创建工作目录。

(2) 编写 Dockerfile 文件。

(3) 生成镜像。

(4) 启动容器测试。

3. 完成任务所需的设备和软件

(1) 一台安装 Windows 10 操作系统的计算机。

(2) VMware Workstation、Docker。

(3) 远程管理工具 MobaXterm。

4. 任务实施步骤

第一步，建立工作目录，操作命令如下：

```
[root@docker ~]# mkdir tomcat
[root@docker ~]# cd tomcat
```

第二步，下载软件包 jdk-8u192-linux-x64.tar.gz 和 apache-tomcat-9.0. 11.tar.gz 并将其上传至工作目录 tomcat 中，操作结果如图 4-13 所示。

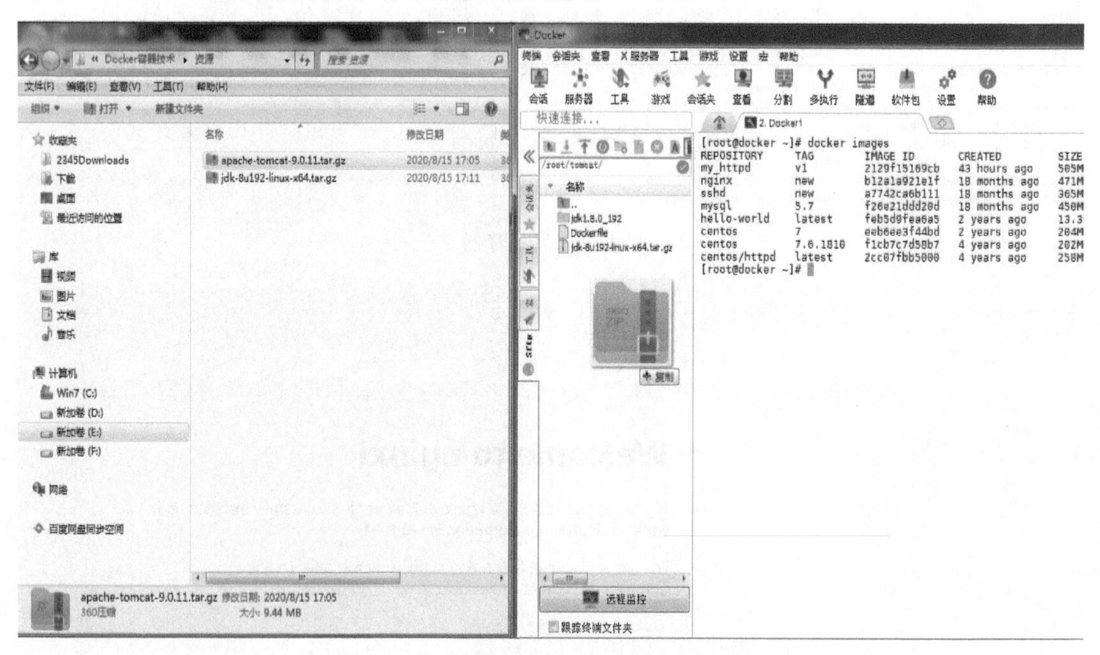

图 4-13　下载所需软件包并上传至工作目录

第三步，解压这两个软件包，并将解压后的 apache 软件包移动到目录 tomc 中，操作命令如下：

```
[root@docker tomcat]# tar xf jdk-8u192-linux-x64.tar.gz
[root@docker tomcat]# tar xf apache-tomcat-9.0.11.tar.gz
[root@docker tomcat]# mv apache-tomcat-9.0.11 tomc
```

第四步，创建并编辑 Dockerfile 文件，操作命令和代码如下：

```
[root@docker tomcat]# vim Dockerfile
```

Dockerfile 文件的内容如下所示，共 6 行代码：

```
1 FROM centos:7
2 MAINTAINER tomcat
3 ENV JAVA_HOME /usr/local/jdk-8u192
4 ADD jdk1.8.0_192 /usr/local/jdk-8u192
5 ADD tomc /usr/local/tomcat
6 EXPOSE 8080
```

第 1 行指明了基础镜像，第 2 行说明镜像维护者的信息，第 3 行设置环境变量，第 4～5

行复制文件到指定位置，第 6 行开启 8080 端口。

第五步，构建镜像并查看，操作命令如下：

```
[root@docker tomcat]# docker build -t tomcat:new .
[root@docker tomcat]# docker images
```

命令运行结果如图 4-14 所示。

```
[root@docker tomcat]# docker build -t tomcat:new .
Sending build context to Docker daemon  614.1MB
Step 1/6 : FROM centos:7
 ---> eeb6ee3f44bd
Step 2/6 : MAINTAINER tomcat
 ---> Running in 04ea2a1d442a
Removing intermediate container 04ea2a1d442a
 ---> b2b6bf178f07
Step 3/6 : ENV JAVA_HOME /usr/local/jdk-8u192
 ---> Running in 193d3db96f64
Removing intermediate container 193d3db96f64
 ---> 630c7642e0ac
Step 4/6 : ADD jdk1.8.0_192 /usr/local/jdk-8u192
 ---> a2b39c143c51
Step 5/6 : ADD tomc /usr/local/tomcat
 ---> 001f05cc390e
Step 6/6 : EXPOSE 8080
 ---> Running in 03c3999703a7
Removing intermediate container 03c3999703a7
 ---> fd0b1833593e
Successfully built fd0b1833593e
Successfully tagged tomcat:new
[root@docker tomcat]# docker images
REPOSITORY      TAG         IMAGE ID        CREATED           SIZE
tomcat          new         fd0b1833593e    32 seconds ago    615MB
my_httpd        v1          2129f15169cb    2 days ago        505MB
nginx           new         b12a1a921e1f    18 months ago     471MB
sshd            new         a7742ca6b111    18 months ago     365MB
mysql           5.7         f26e21ddd20d    18 months ago     450MB
hello-world     latest      feb5d9fea6a5    2 years ago       13.3kB
centos          7           eeb6ee3f44bd    2 years ago       204MB
centos          7.6.1810    f1cb7c7d58b7    4 years ago       202MB
centos/httpd    latest      2cc07fbb5000    4 years ago       258MB
[root@docker tomcat]# 
```

图 4-14　构建 tomcat 镜像

第六步，运行容器并启动 tomcat，操作命令如下：

```
[root@docker tomcat]# docker run -it -p 8081:8080 tomcat:new bash
[root@80fd52bdc6ff /]# /usr/local/tomcat/bin/catalina.sh start
```

命令运行结果如图 4-15 所示。

```
[root@docker tomcat]# docker run -it -p 8081:8080 tomcat:new bash
[root@80fd52bdc6ff /]# /usr/local/tomcat/bin/catalina.sh start
Using CATALINA_BASE:   /usr/local/tomcat
Using CATALINA_HOME:   /usr/local/tomcat
Using CATALINA_TMPDIR: /usr/local/tomcat/temp
Using JRE_HOME:        /usr/local/jdk-8u192
Using CLASSPATH:       /usr/local/tomcat/bin/bootstrap.jar:/usr/local/tomcat/bin/tomcat-juli.jar
Tomcat started.
[root@80fd52bdc6ff /]# 
```

图 4-15　运行容器并启动 tomcat

第七步，通过浏览器访问地址 192.168.1.10，tomcat 启动页面如图 4-16 所示，可见构建的 tomcat 镜像测试成功。

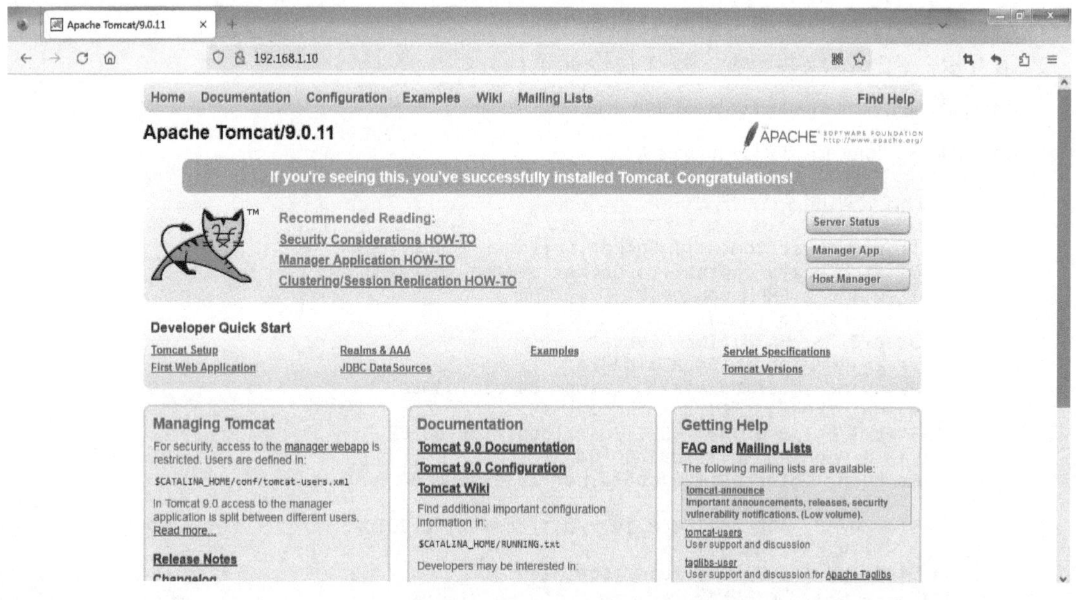

图 4-16　tomcat 启动页面

双创视角

阿里云在淘宝网的应用

淘宝网是亚太地区较大的网络零售商圈，由阿里巴巴集团在 2003 年 5 月创立。阿里云作为淘宝电商主要的云服务提供商，为淘宝提供了大量的计算资源和服务，包括服务器、存储、数据库、网络、安全、大数据分析、机器学习等，确保淘宝电商平台能够高效、稳定地运行，为用户提供良好的购物体验。

阿里云为淘宝提供了丰富的云计算服务，主要包括：

弹性计算服务：帮助淘宝根据业务需求动态调整计算资源，实现灵活的计算能力扩展和回收。

存储服务：为淘宝提供了对象存储 OSS、日志服务 SLS、块存储 EBS、文件存储 NAS 等服务，确保数据的高可用性和可靠性，帮助淘宝实现高效的数据存储和管理，支持大规模用户访问和数据处理。

大数据和人工智能服务：帮助淘宝进行大数据分析，实现智能推荐，提高运营效率和用户体验。

应用服务：Docker 服务帮助淘宝快速部署应用，提高开发效率；OpenStack 服务进行虚拟化管理，提高资源利用率；Kubernetes 服务进行容器化管理，提高部署效率。

安全服务：为淘宝提供了云安全中心、DDoS 防护、数据安全中心等服务，确保平台的安全稳定，帮助淘宝有效应对各种安全威胁，确保用户数据和业务安全。

项 目 小 结

本项目介绍了 Docker 镜像的结构、创建 Docker 镜像的方法、Dockerfile 等知识；完成了通过容器创建镜像、通过Dockerfile构建http、nginx和tomcat镜像等操作任务，让读者掌握创建Docker镜像的基本方法。

习 题 测 试

一、单选题

1. 镜像由(　　)个层组成，每层叠加之后将形成一个独立的对象。

A. 1　　B. 2

C. 4　　D. 多

2. 镜像除了可以在线下载，也可以在本地(　　)镜像。

A. 分享　　B. 链接

C. 复制　　D. 制作

3. Dockerfile把(　　)写成脚本，然后按顺序执行以实现自动创建镜像。

A. 基于镜像执行的命令　　B. 启动容器时运行的命令

C. 创建镜像的步骤　　D. 创建镜像的方法

二、多选题

1. 镜像包含运行某个软件所需的所有内容，比如(　　)等。

A. 代码　　B. 运行所需的库

C. 环境变量　　D. 配置文件

2. 创建镜像的方法通常包括(　　)。

A. 基于容器创建镜像　　B. 基于模板创建镜像

C. 基于Dockerfile创建镜像　　D. 基于仓库创建镜像

3. Dockerfile操作指令VOLUME用于实现挂载，可以将(　　)挂载到(　　)中，完成持久化存储数据。

A. 虚拟机目录　　B. 宿主机目录

C. 镜像　　D. 容器

三、简答题

1. 简述Docker镜像的结构。

2. 简述通过Dockerfile创建镜像的步骤。

项目五
编排容器 Docker Compose

学习目标

(1) 了解 Docker 编排容器。
(2) 理解 Docker Compose 的使用。
(3) 掌握安装 Docker Compose 的方法。
(4) 掌握使用 Docker Compose 部署服务的方法。

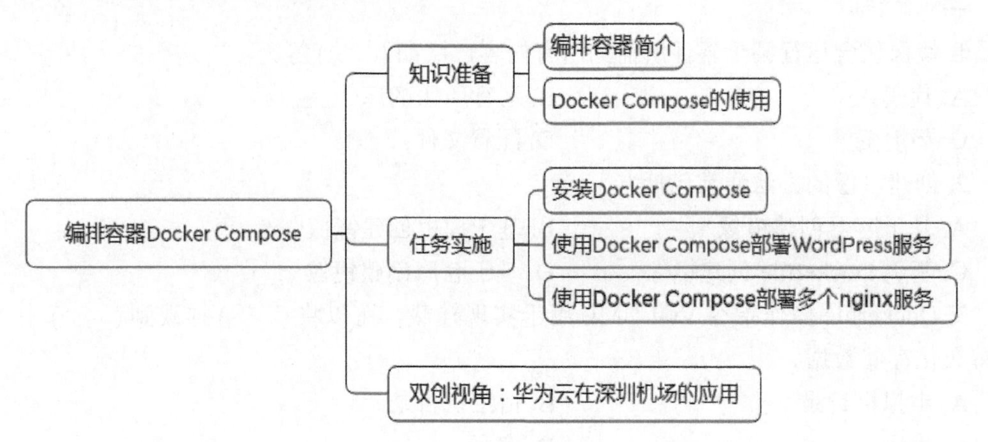

5.1 知识准备

5.1.1 编排容器简介

在现代应用开发中，容器化已经成为一种流行的部署方式，而 Docker 是其中应用最广泛的容器化技术。一个项目中往往需要运行多个 Docker 容器，如果分别构建和启动这些容器就会比较麻烦且不易于管理，反之对所有容器进行统一编排和部署，则会提升项目的开发和运行效率。例如，在创建一个网站时使用 Docker 部署应用程序，需要分别为应用、数据库、nginx 等运行独立的 Docker 容器，如果将这些 Docker 容器放在一起进行统一管理，则有利于该网站的快速创建和高效运行。

为了方便容器化应用程序的部署、扩展和管理，解决手动管理耗时费力的问题，Docker 容器编排技术应运而生，它的作用是管理和协调多个 Docker 容器，通过管理一组容器来提供应用程序的可靠启动、扩展和管理。现代应用由多个服务组成，每个服务可以运行在多个容器实例中，容器编排工具可以自动化部署和管理这些容器实例。

编排容器的工具具有自动扩展容器、管理容器的网络环境、公开和保护 API、监视和管理容器健康状况等重要功能，目前常用的容器编排工具有 Docker Swarm 和 Kubernetes 等。Swarm 是由 Docker 公司推出的开源集群管理平台，Kubernetes 是 Google 推出的开源集群管理系统 (简称 K8s)，Kubernetes 强大的自动化处理能力、便于扩展和高可用性使其得到了广泛的应用。

具体而言，编排容器的工具可以自动完成以下任务：

(1) 自动部署，根据设定的规则将应用程序自动部署到容器集群中。

(2) 弹性伸缩，根据需求自动增加或减少容器的数量。

(3) 服务发现，为容器提供网络连接和服务发现功能。

(4) 资源管理，给资源分配不同的容器，实现负载均衡。

(5) 故障恢复，监控容器状态，当容器出现故障或不可用时进行自动恢复，确保容器正常运行。

Docker Compose 是定义和运行多容器 Docker 应用程序的工具，用户可以使用 YAML 文件来配置应用程序需要的所有服务，它适用于本地开发和测试环境或者小规模的生产环境部署。

5.1.2 Docker Compose 的使用

Compose 项目是 Docker 官方的开源项目，负责实现对 Docker 容器集群的快速编排。通过 Dockerfile 模板文件，用户可以很方便地定义一个单独的应用容器，但是当需要多个容器相互配合来完成某项任务时，Compose 便能发挥优势了。

一个应用在使用 Docker 容器实现时，通常由多个容器组成，开发者可以使用 YAML

文件来配置应用程序需要的所有服务。Compose 通过一个 YAML 配置文件来管理多个 Docker 容器，所有的容器通过 services 来定义，使用 docker-compose 脚本来启动、停止和重启应用中的服务以及所有依赖服务的容器，适用于组合多个容器进行开发的场景。

1. 使用 Compose 的三个步骤

(1) 使用 Dockerfile 定义应用程序的环境。

(2) 使用 docker-compose.yml 定义构成应用程序的服务，这些服务可以在隔离的环境中一起运行。

(3) 执行 docker-compose up 命令来启动并运行整个应用程序。

Compose 中有两个重要的概念——服务 (service) 和项目 (project)。服务即一个应用的容器，可以包括若干运行相同镜像的容器实例；项目即由一组关联的应用容器组成的一个完整业务单元，项目在 docker-compose.yml 文件中定义。

Compose 项目由 Python 编写，通过调用 Docker 服务提供的 API 来对容器进行管理。

2. 编写 YAML 文件的注意事项

YAML 是一种标记语言，可读性强，用来表达数据序列化的格式。YAML 通过缩进表达数据结构，使用空白字符和分行来定义数据的层级关系，相同层次结构的元素左侧对齐。

在编写 YAML 文件时，要注意以下事项：

(1) 严格区分大小写字母。

(2) 使用空格缩进，不能使用 Tab 键。

(3) 缩进的空格数不重要，相同层级的元素左侧对齐即可。

(4) 符号“#”表示注释。

(5) key: value 键值对中的冒号后面必须要有空格。

(6) YAML 文件扩展名为 .yaml 或 .yml。

3. Compose 配置文件常用字段

Compose 配置文件中，使用 version、services、networks 和 volumes 将其分为四个部分，其中 version 指定 Compose 配置文件的版本，services 定义服务，networks 设置网络，volumes 定义数据卷。Compose 配置文件的常用字段及其含义见表 5-1 所示。

表 5-1 Compose 配置文件常用字段及其含义

字 段	含 义
build	在构建时应用的配置项
context	指定包含 Dockerfile 的目录路径或 git 仓库 url
dockerfile	指定 Dockerfile 文件构建镜像
image	指定启动容器的镜像
command	覆盖容器启动后默认执行的命令
container_name	指定自定义容器的名称，而不使用默认名称
hostname	设置容器的主机名，容器可以通过主机名来相互访问，即使它们位于不同的容器之中

续表

字　段	含　　义
deploy	指定部署和运行服务的相关配置，仅在 swarm mode 下生效
depends_on	指定服务之间的依赖关系，解决服务启动先后顺序的问题
ports	暴露容器端口
links	在不同容器之间创建网络连接
volumes	指定所挂载的主机路径或数据卷名称
networks	设置容器网络连接以获取构建过程中的 RUN 指令
environment	设置环境变量

4. Docker-Compose 文件结构

下面是 Docker-Compose 文件示例，从中可以看到 Docker-Compose 文件的结构。

```
version: '3'
services:
  nginx:
    hostname: nginx
    build:
      context: ./nginx
      dockerfile: Dockerfile
    ports:
      - 81:80
    links:
      - php:php-cgi
    volumes:
      - ./wwwroot:/usr/local/nginx/html

  mysql:
    hostname: mysql
    image: mysql:5.6
    ports:
      - 3306:3306
    volumes:
      - ./mysql/conf:/etc/mysql/conf.d
      - ./mysql/data:/var/lib/mysql
    environment:
      MYSQL_ROOT_PASSWORD: 123456
      MYSQL_USER: user
      MYSQL_PASSWORD: user123
```

5. Docker-Compose 命令格式

```
docker-compose [-f <arg>…] [options] COMMAND [ARGS…]
```

options 选项及其说明见表 5-2 所示。

表 5-2 选项及其说明

选 项	说 明
-f	指定 Compose 模板文件，默认为 docker-compose.yml，可以多次指定
-p	指定项目名称，默认将正在使用的所在目录名称作为项目名
-verbose	输出更多调试信息
-v	打印版本并退出

COMMAND 命令及其说明见表 5-3 所示。

表 5-3 命令及其说明

命 令	说 明
build	重新构建服务
up	构建镜像、创建并启动服务、关联服务相关容器等
down	停止启动的容器，并移除网络
exec	进入指定的容器
ps	列出项目中目前的所有容器
rm	删除所有 (停止状态的) 服务容器
top	查看各个服务容器内运行的进程
logs	查看容器输出
images	显示所有镜像
start/ stop/ restart	启动 / 停止 / 重启服务容器

5.2 任务实施

5.2.1 安装 Docker Compose

安装 Docker Compose

1. 任务目标

掌握安装 Docker Compose 的方法。

2. 任务内容

(1) 安装 Docker Compose。

(2) 赋予 docker compose 文件执行权限。

3. 完成任务所需的设备和软件

(1) 一台安装 Windows 10 操作系统的计算机。

(2) VMware Workstation、Docker、Docker Compose。

(3) 远程管理工具 MobaXterm。

4. 任务实施步骤

第一步，安装 Docker Compose，操作命令如下：

```
[root@docker ~]# curl -L https://github.com/docker/compose/releases/download/v2.23.3/docker-compose-`uname -s`-`uname -m` -o /usr/local/bin/docker-compose
```

其中：

-L：让 HTTP 请求跟随服务器的重定向，curl 默认不跟随重定向。

-o：指定输出文件的位置和文件名。

uname -s：显示内核名称，用于区分不同的操作系统。

uname -m：显示机器的硬件 (CPU) 架构。

第二步，赋予 docker compose 文件执行权限，操作命令如下：

```
[root@docker ~]# chmod +x /usr/local/bin/docker-compose
```

第三步，查看 Docker Compose 版本，操作命令如下：

```
[root@docker ~]# docker-compose -v
```

命令运行结果如图 5-1 所示。

```
[root@docker ~]# curl -L https://github.com/docker/compose/releases/download/v2.23.3/docker-compose-`uname -s`-`uname -m` -o /usr/local/bi
n/docker-compose
  % Total    % Received % Xferd  Average Speed   Time    Time     Time  Current
                                 Dload  Upload   Total   Spent    Left  Speed
  0     0    0     0    0     0      0      0 --:--:--  0:00:01 --:--:--     0
100 56.9M  100 56.9M    0     0   113k      0  0:08:33  0:08:33 --:--:--  177k
[root@docker ~]# docker-compose -v
-bash: /usr/local/bin/docker-compose: Permission denied
[root@docker ~]# chmod +x /usr/local/bin/docker-compose
[root@docker ~]# docker-compose -v
Docker Compose version v2.23.3
```

图 5-1　查看 Docker Compose 版本

5.2.2　使用 Docker Compose 部署 WordPress 服务

使用 Docker Compose 部署 WordPress 服务

1. 任务目标

掌握使用 Docker Compose 部署 WordPress 服务的方法。

2. 任务内容

(1) 创建项目文件夹。

(2) 创建并编辑配置文件 docker-compos.yml。

(3) 创建和启动 WordPress 服务。

(4) 查看容器是否启动。

3. 完成任务所需的设备和软件

(1) 一台安装 Windows 10 操作系统的计算机。

(2) VMware Workstation，Docker，Docker Compose。

(3) 远程管理工具 MobaXterm。

4. 任务实施步骤

第一步，创建项目文件夹并进入其中，操作命令如下：

```
[root@docker ~]# mkdir wordpress
[root@docker ~]# cd wordpress
```

第二步，创建并编辑配置文件 docker-compos.yml，操作命令如下：

```
[root@docker wordpress]# vim docker-compose.yml
```

docker-compose.yml 文件的内容如下：

```
1 version: '3'
2 services:
3   db:
4     image: mariadb
5     volumes:
6       - db_data:/var/lib/mysql
7     restart: always
8     environment:
9       MYSQL_ROOT_PASSWORD: example
10  wordpress:
11    depends_on:
12      - db
13    image: wordpress:4.9.4
14    ports:
15      - "8000:80"
16    restart: always
17    environment:
18      WORDPRESS_DB_HOST: db:3306
19      WORDPRESS_DB_USER: root
20      WORDPRESS_DB_PASSWORD: example
21      WORDPRESS_DB_NAME: wordpress
22 volumes:
23   db_data:
```

第三步，创建和启动 WordPress 服务，操作命令如下：

```
[root@docker wordpress]# docker-compose up -d
```

命令运行结果如图 5-2 所示。

```
[root@docker wordpress]# docker-compose up -d
Pulling db (mariadb:)...
latest: Pulling from library/mariadb
43f89b94cd7d: Pull complete
dfb413a01c7e: Pull complete
1d3f76b535d3: Pull complete
f7efd05ec01e: Pull complete
fe2ff83c75df: Pull complete
50ee0c078c93: Pull complete
6975e72928bb: Pull complete
561d1b426cbd: Pull complete
Digest: sha256:2403cc521634162f743b5179ff5b35520daf72df5d9e7e397192af685d9148fd
Status: Downloaded newer image for mariadb:latest
Creating wordpress_db_1 ... done
Creating wordpress_wordpress_1 ... done
```

图 5-2　创建和启动 WordPress 服务

第四步，查看容器是否启动，操作命令如下：

```
[root@docker wordpress]# docker-compose ps
```

命令运行结果如图 5-3 所示，可见两个容器已经启动起来。

```
[root@docker wordpress]# docker-compose ps
        Name                      Command               State                  Ports
------------------------------------------------------------------------------------------------------
wordpress_db_1          docker-entrypoint.sh mariadbd   Up      3306/tcp
wordpress_wordpress_1   docker-entrypoint.sh apach ...  Up      0.0.0.0:8000->80/tcp,:::8000->80/tcp
```

图 5-3　查看容器

第五步，查看容器分别使用的镜像，操作命令如下：

```
[root@docker wordpress]# docker-compose images
```

命令运行结果如图 5-4 所示，可见两个容器使用的镜像分别为 mariadb:latest 和 wordpress:4.9.4。

```
[root@docker wordpress]# docker-compose images
      Container           Repository    Tag       Image Id       Size
----------------------------------------------------------------------
wordpress_db_1            mariadb      latest   f35870862d64   403.9 MB
wordpress_wordpress_1     wordpress    4.9.4    e6bc04e5d2ab   441.9 MB
```

图 5-4　查看容器分别使用的镜像

第六步，在浏览器中输入地址 http://192.168.1.10:8000，访问 wordpress 页面，如图 5-5 所示。

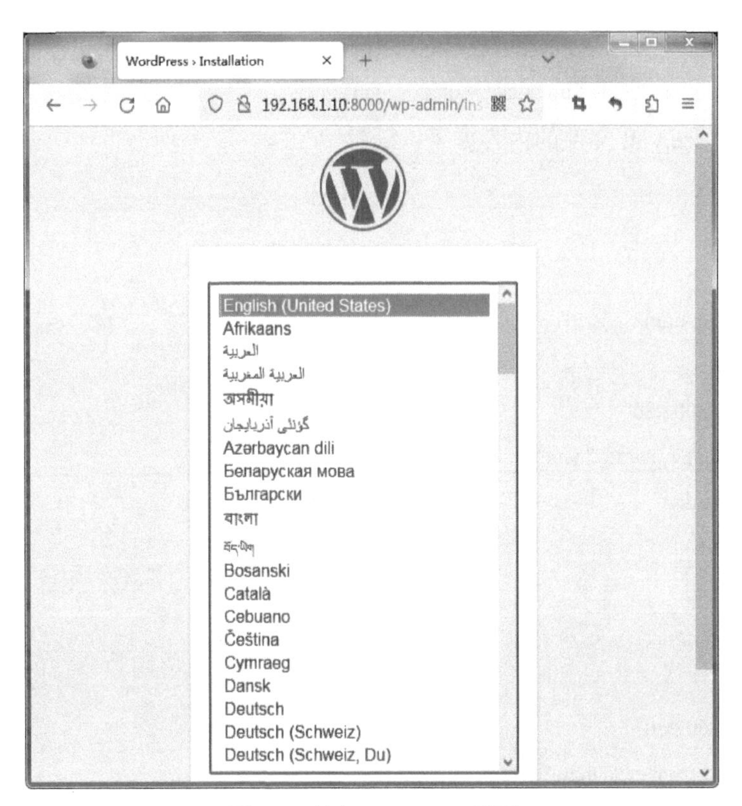

图 5-5　访问 WordPress 页面

5.2.3 使用 Docker Compose 部署多个 nginx 服务

使用 Docker Compose 部署多个 nginx 服务

1. 任务目标

掌握使用 Docker Compose 部署多个 nginx 服务的方法。

2. 任务内容

(1) 创建项目文件夹。

(2) 创建并编辑配置文件 docker-compose.yml。

(3) 创建和启动多个 nginx 服务。

(4) 查看容器是否启动。

3. 完成任务所需的设备和软件

(1) 一台安装 Windows 10 操作系统的计算机。

(2) VMware Workstation、Docker、Docker Compose。

(3) 远程管理工具 MobaXterm。

4. 任务实施步骤

第一步，创建项目文件夹并进入其中，操作命令如下：

```
[root@docker ~]# mkdir my_nginxs
[root@docker ~]# cd my_nginxs
```

第二步，创建并编辑配置文件 docker-compose.yml，操作命令如下：

```
[root@docker my_nginxs]# vim docker-compose.yml
```

docker-compose.yml 文件的内容如下：

```
1  version: '3'
2  services:
3    neta:
4      image: nginx
5      ports:
6        - "8001:80"
7      container_name: "neta"
8      networks:
9        - nw-1
10   netb:
11     image: nginx
12     ports:
13       - "8002:80"
14     container_name: "netb"
15     networks:
```

```
16         - nw-2
17     netc:
18       image: nginx
19       ports:
20         - "8003:80"
21       container_name: "netc"
22       networks:
23         - nw-3
24     netd:
25       image: nginx
26       ports:
27         - "8004:80"
28       container_name: "netd"
29       networks:
30         - nw-4
31     nete:
32       image: nginx
33       ports:
34        - "8005:80"
35      container_name: "nete"
36      networks:
37        - nw-1
38        - nw-2
39        - nw-3
40        - nw-4
41 networks:
42   nw-1:
43     driver: bridge
44   nw-2:
45     driver: bridge
46   nw-3:
47     driver: bridge
48   nw-4:
49     driver: bridge
```

第三步，创建和启动多个 nginx 服务，操作命令如下：

```
[root@docker my_nginxs]# docker-compose up -d
```

命令运行结果如图 5-6 所示。

```
[root@docker my_nginxs]# docker-compose up -d
Creating network "my_nginxs_nw-1" with driver "bridge"
Creating network "my_nginxs_nw-2" with driver "bridge"
Creating network "my_nginxs_nw-3" with driver "bridge"
Creating network "my_nginxs_nw-4" with driver "bridge"
Pulling neta (nginx:)...
latest: Pulling from library/nginx
578acb154839: Pull complete
e398db710407: Pull complete
85c41ebe6d66: Pull complete
7170a263b582: Pull complete
8f28d06e2e2e: Pull complete
6f837de2f887: Pull complete
c1dfc7e1671e: Pull complete
Digest: sha256:86e53c4c16a6a276b204b0fd3a8143d86547c967dc8258b3d47c3a21bb68d3c6
Status: Downloaded newer image for nginx:latest
Creating netd ... done
Creating nete ... done
Creating netc ... done
Creating netb ... done
Creating neta ... done
```

图 5-6　创建和启动多个 nginx 服务

第四步，查看容器是否启动，操作命令如下：

```
[root@docker my_nginxs]# docker-compose ps
```

命令运行结果如图 5-7 所示，可见五个容器已经启动起来。

```
[root@client my_nginxs]# docker-compose ps
Name        Command               State                  Ports
-------------------------------------------------------------------------------------
neta   /docker-entrypoint.sh ngin ...   Up      0.0.0.0:8001->80/tcp,:::8001->80/tcp
netb   /docker-entrypoint.sh ngin ...   Up      0.0.0.0:8002->80/tcp,:::8002->80/tcp
netc   /docker-entrypoint.sh ngin ...   Up      0.0.0.0:8003->80/tcp,:::8003->80/tcp
netd   /docker-entrypoint.sh ngin ...   Up      0.0.0.0:8004->80/tcp,:::8004->80/tcp
nete   /docker-entrypoint.sh ngin ...   Up      0.0.0.0:8005->80/tcp,:::8005->80/tcp
[root@client my_nginxs]#
```

图 5-7　查看容器

第五步，查看容器分别使用的镜像，操作命令如下：

```
[root@docker my_nginxs]# docker-compose images
```

命令运行结果如图 5-8 所示，可见五个容器使用的镜像均为 nginx:latest。

```
[root@client my_nginxs]# docker-compose images
Container   Repository    Tag       Image Id       Size
---------------------------------------------------------
neta        nginx        latest   c20060033e06   178 MB
netb        nginx        latest   c20060033e06   178 MB
netc        nginx        latest   c20060033e06   178 MB
netd        nginx        latest   c20060033e06   178 MB
nete        nginx        latest   c20060033e06   178 MB
[root@client my_nginxs]#
```

图 5-8　查看容器使用的镜像

第六步，在浏览器中输入地址 http://192.168.1.10:8001(或 8002 或 8003 或 8004 或 8005) 可以访问 nginx 页面，如图 5-9 所示。

图 5-9　访问 nginx 页面

双创视角

华为云在深圳机场的应用

深圳宝安国际机场是中国境内集海、陆、空、铁联运为一体的现代化大型国际空港。作为唯一加入国际航空运输协会(IATA)“未来机场”项目的内地机场，2017 年，深圳机场集团与华为签订战略合作协议，并于 2018 年开始在国内机场中率先全面、系统地启动数字化转型。

华为跟深圳机场深度战略合作“未来机场”项目，联合打造“机场智能体”，提供从云平台、视频监控、集成平台和 AI 应用算法在内的全栈解决方案。华为提供“企业服务总线”ROMA 平台，接入各个新建系统和机场原有业务系统，在保持机场现有业务架构不破坏的情况下实现了运控管理等业务的智能化。

华为云为深圳机场针对“运控一张图”“出行一张脸”的场景提供了多个 AI 应用，如机位自动分配、刷脸快速安检等，算法由华为和合作方提供，基于华为云平台开发和训练。

项 目 小 结

本项目介绍了编排容器的概念及其要完成的任务，以及 Docker Compose 的基本使用方法等知识；完成了安装 Docker Compose、使用 Docker Compose 部署 WordPress 服务及多个 nginx 服务等操作任务，让读者学会使用 Docker Compose 进行容器编排的基本方法。

习 题 测 试

一、单选题

1. 一个项目中往往需要运行(　　)Docker 容器。

A. 1 个　　　　B. 2 个

C. 3 个　　　　D. 多个

2. (　　)技术的作用是管理和协调多个 Docker 容器，通过管理一组容器来提供应用程序的可靠启动、扩展和管理。

A. Docker 镜像管理　　B. Docker 容器编排

C. Dockerfile　　D. Docker 网络通信

3. Docker Compose 使用 (　　) 文件定义构成应用程序的服务，这些服务可以在隔离的环境中一起运行。

A. 镜像　　B. 容器

C. docker-compose.yml　　D. Dockerfile

二、多选题

1. 容器编排工具具有 (　　) 等重要功能。

A. 自动化地扩展容器　　B. 管理容器的网络环境

C. 公开和保护 API　　D. 监视和管理容器健康状况

2. Docker Compose 是定义和运行多容器 Docker 应用程序的工具，用户可以使用 YAML 文件来配置应用程序需要的所有服务，它适用于 (　　)。

A. 本地开发　　B. 测试环境

C. 小规模的生产环境部署　　D. 制作镜像

3. Compose 配置文件中，使用 (　　) 将其分为四个部分。

A. version　　B. services

C. networks　　D. volumes

三、简答题

1. 简述容器编排工具可以自动化地完成哪些任务。

2. 简述使用 Docker Compose 进行容器编排的步骤。

项目六
部署和管理 Harbor 私有仓库

学习目标

(1) 了解公有仓库。
(2) 了解 Harbor 私有仓库。
(3) 掌握部署 Harbor 私有仓库的方法。
(4) 掌握管理 Harbor 私有仓库的方法。

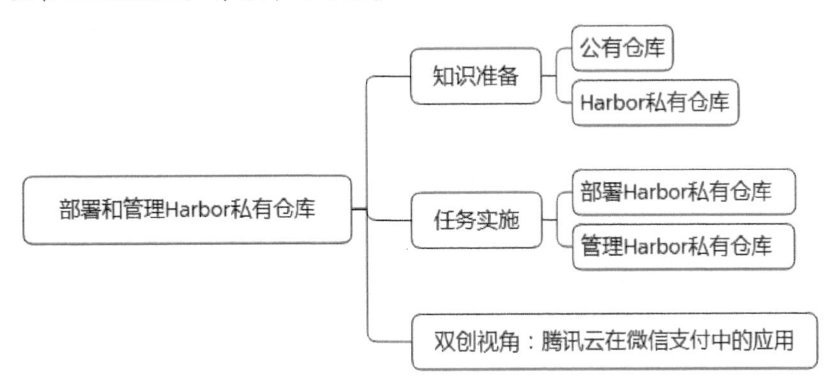

6.1 知识准备

6.1.1 公有仓库

Docker 的三大基础组件是镜像、容器和仓库。Docker 官方给用户提供了一个 Docker 仓库，它就像手机里的 APP 应用市场一样，里面存放着各种各样已经打包好的应用——Docker 镜像。用户可以将需要的 Docker 镜像下载到本地，然后基于 Docker 镜像创建 Docker 容器，容器会启动预先定义好的进程与用户交互，对外提供服务。

Docker 仓库用于镜像的存储，是 Docker 镜像分发和部署的关键。在实际应用的过程中，由开发者或者运维制作好的应用程序镜像，可以上传到镜像仓库。我们可以使用官方的公有仓库 Docker Hub，也可以搭建自己的私有仓库，Docker 运行中使用的默认仓库就是 Docker Hub 公共仓库。

Docker Hub 是 Docker 官方维护的一个公有镜像仓库，它积累了大量的官方和社区创建的镜像，是目前最大的一个公有仓库，用户可以直接搜索并使用其中的镜像。在上传和分享自己创建的镜像时，需要申请账户，首先通过网址 https://hub.docker.com 登陆 Web 界面，完成注册之后就可以使用自己的用户名登录了。

有了 Docker Hub 的账户之后，就可以利用命令“docker login”登录 Docker Hub，然后利用命令“docker push 镜像名称：版本”上传镜像，利用命令“docker search 镜像名称”搜索镜像，利用命令“docker pull 镜像名称”下载镜像。

6.1.2 Harbor 私有仓库

我们发现，本地访问 Docker Hub 的速度往往是很慢的，而且在生产环境中，企业通常会创建符合实际需求的镜像，考虑到安全问题，这些镜像必须存放在企业内部，并且确保不受网络限制，可以快速地下载或上传镜像。这些情况下，个人或企业就要搭建自己的私有仓库，仅供个人或企业内部使用。

使用 Docker Hub 中的仓库镜像时，首先下载仓库镜像 registry 到本地，然后运行仓库镜像 registry 为容器，该容器即可为用户提供上传镜像、拉取镜像等私有仓库服务。但这种方式的镜像保存在容器中，一旦容器被删除后，私有仓库及其中的镜像也被一并删除，因此安全系数较低。

Harbor 是 VMware 公司开源的构建企业级私有 docker 镜像仓库的解决方案，它以 Docker 公司开源的 Registry 为基础，提供了友好的 Web UI 界面、角色和用户权限管理、用户操作审计等功能，是 Docker Registry 的更高级封装。

1. Harbor 的特性

(1) 基于角色控制：用户和仓库都是基于项目进行组织，用户在项目中可以拥有不同的权限。

(2) 基于镜像的复制策略：镜像可以在多个 Harbor 实例之间复制 (同步)，适用于负载均衡、高可用性、多数据中心、混合和多云的场景。

(3) 支持 LDAP/AD：Harbor 可以集成企业 AD/LDAP 进行用户认证和管理。

(4) 镜像删除和空间回收：镜像可以删除并回收镜像占用的空间。

(5) 图形化 UI：用户可以通过浏览器浏览、搜索镜像仓库以及对项目进行管理。

(6) 审计管理：对镜像仓库的所有操作进行记录。

(7) 支持 RESTful API：易与其他管理软件进行集成，提供更多对于 Harbor 的管理操作。

2. Harbor 的构成

Harbor 在架构上主要有 Proxy、Registry、Core services、Database(Harbor-db)、Log collector (Harbor-log)、Job service 六个组件，Harbor 的架构图如图 6-1 所示。

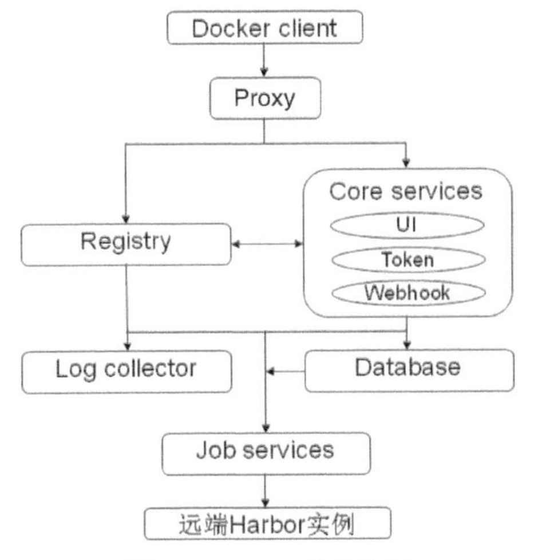

图 6-1　Harbor 的架构图

(1) Proxy：nginx 的前段代理，将来自浏览器、客户端的请求转发到后端不同的服务上，如 Harbor 的 Registry、UI、Token 服务等。

(2) Registry：负责存储 Docker 镜像，并处理 Docker push/pull 命令。由于要对用户进行访问控制，即不同用户对 Docker 镜像有不同的读写权限，Registry 会指向一个 Token 服务，强制用户每次 Docker pull/push 请求都要携带一个合法的 Token，Registry 会通过公钥对 Token 进行解密验证。

(3) Core services：Harbor 的核心功能，主要提供的服务如下。

UI：图形化界面，用于管理 Registry 上的镜像和用户权限。

Token：根据用户权限给每个 Docker push/pull 命令签发通行证。Docker 客户端向 Registry 服务发起的请求如果不包含 Token，会被重定向，获得 Token 后再重新向 Registry

发起请求。

Webhook：在 Registry 上配置 Webhook，把镜像状态变化情况传递给 UI 模块。

(4) Database：提供数据库服务，其中存储用户权限、审计日志、Docker 镜像分组信息等数据。

(5) Log collector：监控 Harbor 运行，收集其他组件的日志。

(6) Job services：主要用于镜像复制，本地镜像可以被同步到远程 Harbor 实例上。

Harbor 的每个组建均以 Docker 容器形式构建，因此使用 Docker Compose 对其进行部署，可以在 docker-compose.yml 所在目录中执行 docker-compose ps 命令查看各个运行的容器。

6.2 任务实施

6.2.1 部署 Harbor 私有仓库

部署 Harbor 私有仓库

1. 任务目标

掌握部署 Harbor 私有仓库的方法。

2. 任务内容

安装与启动 Harbor 服务。

3. 完成任务所需的设备和软件

(1) 一台安装 Windows 10 操作系统的计算机。

(2) VMware Workstation、Docker、Docker Compose。

(3) 远程管理工具 MobaXterm。

4. 任务实施步骤

第一步，安装 Docker Compose(如果已经安装，此处第一步和第二步可以忽略)，操作命令如下：

```
[root@docker ~]# curl -L https://github.com/docker/compose/releases/download/v2.23.3/docker-compose-`uname -s`-`uname -m` -o /usr/local/bin/docker-compose
```

第二步，添加 docker compose 文件执行权限，操作命令如下：

```
[root@docker ~]# chmod +x /usr/local/bin/docker-compose
```

第三步，查看 Docker Compose 版本，操作命令如下：

```
[root@docker ~]# docker-compose -v
```

命令运行结果如图 6-2 所示。

```
[root@docker ~]# curl -L https://github.com/docker/compose/releases/download/v2.23.3/docker-compose-`uname -s`-`uname -m` -o /usr/local/bi
n/docker-compose
  % Total    % Received % Xferd  Average Speed   Time    Time     Time  Current
                                 Dload  Upload   Total   Spent    Left  Speed
  0     0    0     0    0     0      0      0 --:--:--  0:00:01 --:--:--     0
100 56.9M  100 56.9M    0     0   113k      0  0:08:33  0:08:33 --:--:--  177k
[root@docker ~]# docker-compose -v
-bash: /usr/local/bin/docker-compose: Permission denied
[root@docker ~]# chmod +x /usr/local/bin/docker-compose
[root@docker ~]# docker-compose -v
Docker Compose version v2.23.3
```

图 6-2　查看 Docker Compose 版本

第四步，将附带资源中的软件包 harbor-offline-installer-v2.9.1.tgz 拖拽至虚拟机 /usr/local/ 目录下，操作过程如图 6-3 所示。

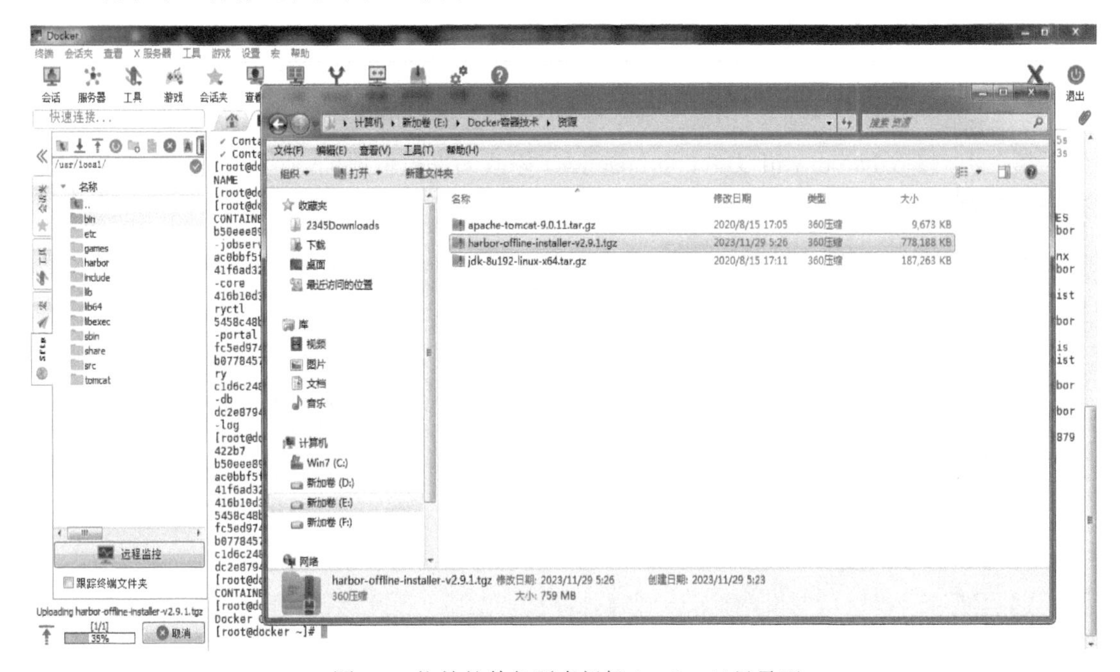

图 6-3　拖拽软件包至虚拟机 /usr/local/ 目录下

第五步，将软件包进行解压，操作命令如下：

```
[root@docker ~]# cd /usr/local
[root@docker local]# ls
[root@docker local]# tar xvf harbor-offline-installer-v2.9.1.tgz
[root@docker local]# ls
```

命令运行结果如图 6-4 所示。

```
[root@docker ~]# cd /usr/local
[root@docker local]# ls
bin  etc  games  harbor-offline-installer-v2.9.1.tgz  include  lib  lib64  libexec  sbin  share  src  tomcat
[root@docker local]# tar xvf harbor-offline-installer-v2.9.1.tgz
harbor/harbor.v2.9.1.tar.gz
harbor/prepare
harbor/LICENSE
harbor/install.sh
harbor/common.sh
harbor/harbor.yml.tmpl
[root@docker local]# ls
bin  etc  games  harbor  harbor-offline-installer-v2.9.1.tgz  include  lib  lib64  libexec  sbin  share  src  tomcat
[root@docker local]#
```

图 6-4　解压软件包

第六步，进入 harbor 目录，复制 harbor.yml.tmpl 文件为 harbor.yml 文件，操作命令如下：

```
[root@docker local]# cd harbor
[root@docker harbor]# ls
[root@docker harbor]# cp harbor.yml.tmpl harbor.yml
[root@docker harbor]# ls
```

命令运行结果如图 6-5 所示。

```
[root@docker local]# cd harbor
[root@docker harbor]# ls
common.sh  harbor.v2.9.1.tar.gz  harbor.yml.tmpl  install.sh  LICENSE  prepare
[root@docker harbor]# cp harbor.yml.tmpl harbor.yml
[root@docker harbor]# ls
common.sh  harbor.v2.9.1.tar.gz  harbor.yml  harbor.yml.tmpl  install.sh  LICENSE  prepare
```

图 6-5　复制 harbor.yml.tmpl 文件为 harbor.yml 文件

第七步，修改 harbor.yml 文件，将 hostname 修改为虚拟机地址，并将 https 协议注释掉，操作命令如下：

```
[[root@docker harbor]# vim harbor.yml
```

文件内容修改情况如图 6-6 所示。

```
# Configuration file of Harbor

# The IP address or hostname to access admin UI and registry service.
# DO NOT use localhost or 127.0.0.1, because Harbor needs to be accessed by external clients.
hostname: 192.168.1.10

# http related config
http:
  # port for http, default is 80. If https enabled, this port will redirect to https port
  port: 80

# https related config
#https:
  # https port for harbor, default is 443
  # port: 443
  # The path of cert and key files for nginx
  # certificate: /your/certificate/path
  # private_key: /your/private/key/path

# # Uncomment following will enable tls communication between all harbor components
# internal_tls:
#   # set enabled to true means internal tls is enabled
#   enabled: true
#   # put your cert and key files on dir
#   dir: /etc/harbor/tls/internal
#   # enable strong ssl ciphers (default: false)
#   strong_ssl_ciphers: false

# Uncomment external_url if you want to enable external proxy
# And when it enabled the hostname will no longer used
# external_url: https://reg.mydomain.com:8433

# The initial password of Harbor admin
# It only works in first time to install harbor
# Remember Change the admin password from UI after launching Harbor.
harbor_admin_password: Harbor12345

# Harbor DB configuration
"harbor.yml" 306L, 13759C
```

图 6-6　修改 harbor.yml 文件

第八步，启动 harbor，操作命令如下：

```
[root@docker harbor]# sh install.sh
```

命令运行结果如图 6-7 所示。

```
[root@docker harbor]# sh install.sh

[Step 0]: checking if docker is installed ...

Note: docker version: 20.10.14

[Step 1]: checking docker-compose is installed ...

Note: docker-compose version: 2.23.3

[Step 2]: loading Harbor images ...
Loaded image: goharbor/harbor-jobservice:v2.9.1
Loaded image: goharbor/harbor-registryctl:v2.9.1
Loaded image: goharbor/harbor-core:v2.9.1
Loaded image: goharbor/harbor-log:v2.9.1
Loaded image: goharbor/harbor-db:v2.9.1
Loaded image: goharbor/harbor-exporter:v2.9.1
Loaded image: goharbor/redis-photon:v2.9.1
Loaded image: goharbor/nginx-photon:v2.9.1
Loaded image: goharbor/registry-photon:v2.9.1
Loaded image: goharbor/trivy-adapter-photon:v2.9.1
Loaded image: goharbor/prepare:v2.9.1
Loaded image: goharbor/harbor-portal:v2.9.1


[Step 3]: preparing environment ...

[Step 4]: preparing harbor configs ...
prepare base dir is set to /usr/local/harbor
```

图 6-7　启动 Harbor

第九步，查看 harbor 启动镜像，操作命令如下：

[root@docker harbor]# docker-compose ps

命令运行结果如图 6-8 所示。

```
[root@docker harbor]# docker-compose ps
NAME                IMAGE                                COMMAND                  SERVICE       CREATED          STATUS                    PORTS
harbor-core         goharbor/harbor-core:v2.9.1          "/harbor/entrypoint.…"   core          14 minutes ago   Up 14 minutes (healthy)
harbor-db           goharbor/harbor-db:v2.9.1            "/docker-entrypoint.…"   postgresql    14 minutes ago   Up 14 minutes (healthy)
harbor-jobservice   goharbor/harbor-jobservice:v2.9.1    "/harbor/entrypoint.…"   jobservice    14 minutes ago   Up 14 minutes (healthy)
harbor-log          goharbor/harbor-log:v2.9.1           "/bin/sh -c /usr/loc…"   log           14 minutes ago   Up 14 minutes (healthy)   127.0.0.1:1514->10514/tcp
harbor-portal       goharbor/harbor-portal:v2.9.1        "nginx -g 'daemon of…"   portal        14 minutes ago   Up 14 minutes (healthy)
nginx               goharbor/nginx-photon:v2.9.1         "nginx -g 'daemon of…"   proxy         14 minutes ago   Up 14 minutes (healthy)   0.0.0.0:80->8080/tcp, :::80->8080/tcp
redis               goharbor/redis-photon:v2.9.1         "redis-server /etc/r…"   redis         14 minutes ago   Up 14 minutes (healthy)
registry            goharbor/registry-photon:v2.9.1      "/home/harbor/entryp…"   registry      14 minutes ago   Up 14 minutes (healthy)
registryctl         goharbor/harbor-registryctl:v2.9.1   "/home/harbor/start.…"   registryctl   14 minutes ago   Up 14 minutes (healthy)
```

图 6-8　启动 Harbor

6.2.2　管理 Harbor 私有仓库

管理 Harbor 私有仓库

1. 任务目标

学会管理 Harbor 私有仓库。

2. 任务内容

(1) 通过服务器管理 Harbor 私有仓库。

(2) 通过客户机管理 Harbor 私有仓库。

3. 完成任务所需的设备和软件

(1) 一台安装 Windows 10 操作系统的计算机。

(2) VMware Workstation、Docker、Docker Compose。

(3) 远程管理工具 MobaXterm。

4. 任务实施步骤

第一步，打开浏览器访问网址 http://192.168.1.10，进入 Harbor 私有仓库的登录页面，如

图 6-9 所示。

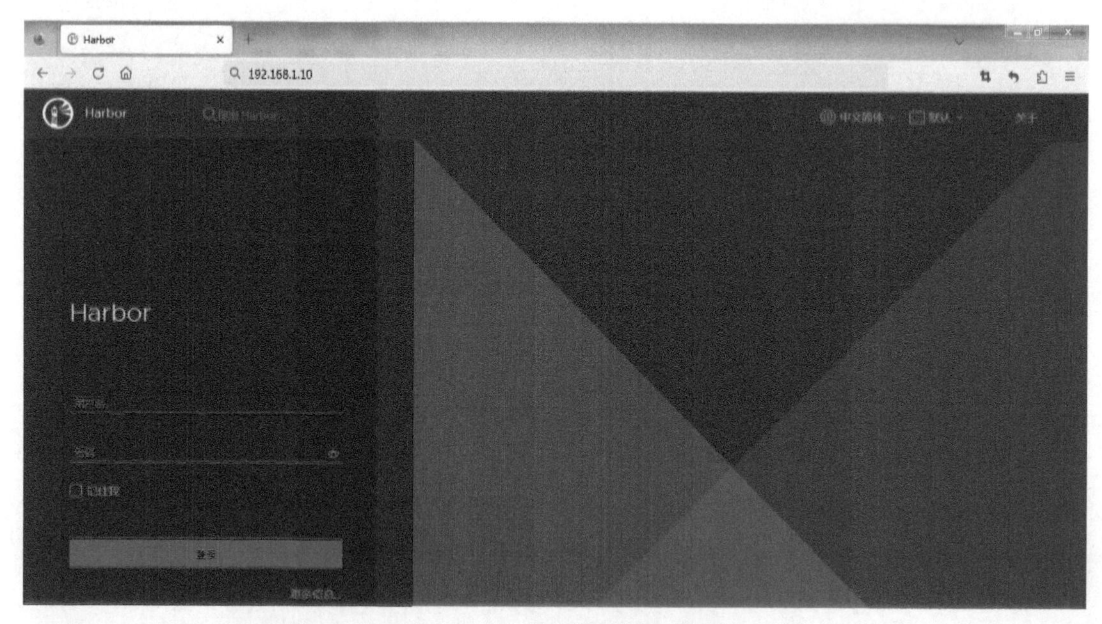

图 6-9　Harbor 私有仓库的登陆页面

第二步，输入默认管理员的用户名 admin 和密码 Harbor12345，进入 Harbor 私有仓库的管理页面，如图 6-10 所示。所有基础镜像都会放在 library 里面，这是一个公开的镜像仓库。

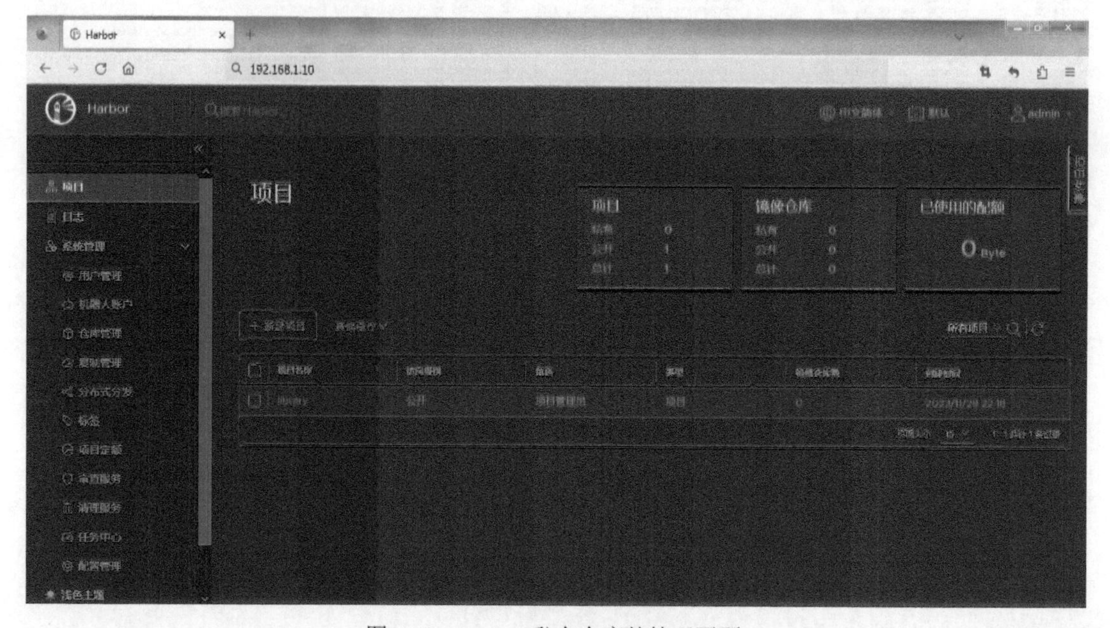

图 6-10　Harbor 私有仓库的管理页面

第三步，单击“+ 新建项目”按钮，打开新建项目对话框，如图 6-11 所示。输入项目名称 mytest，勾选访问级别的“公开”即为公开项目，不勾选即为私有项目。对于共有仓库，不需要执行 docker login 即可下载镜像。

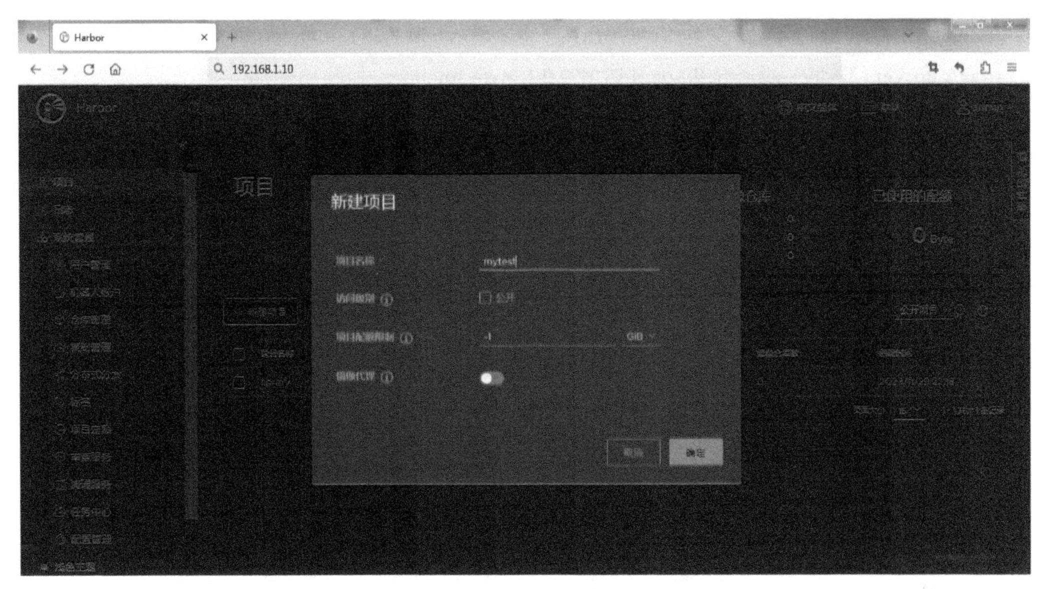

图 6-11　新建项目

第四步，单击“确定”按钮，即可成功创建新项目，如图 6-12 所示。

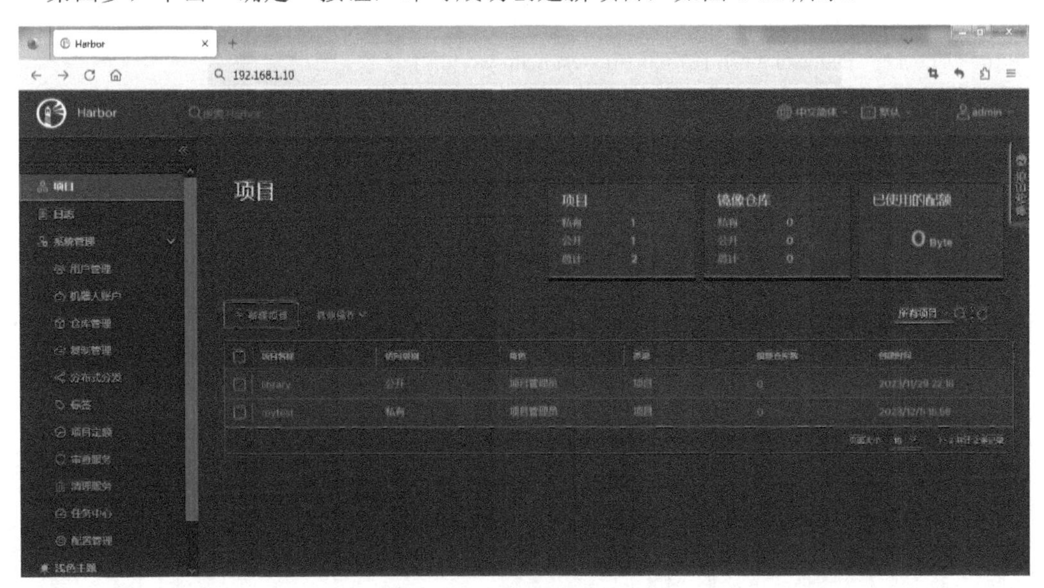

图 6-12　成功创建新项目

第五步，在本地登录 Harbor，操作命令如下：

```
[root@docker ~]# docker login -u admin -p Harbor12345 http://127.0.0.1
```

命令运行结果如图 6-13 所示。

```
[root@docker ~]# docker login -u admin -p Harbor12345 http://127.0.0.1
WARNING! Using --password via the CLI is insecure. Use --password-stdin.
WARNING! Your password will be stored unencrypted in /root/.docker/config.json.
Configure a credential helper to remove this warning. See
https://docs.docker.com/engine/reference/commandline/login/#credentials-store

Login Succeeded
```

图 6-13　在本地登录 Harbor

第六步，将本地已有镜像 hello-world 打上标签，操作命令如下：

```
[root@docker ~]# docker tag hello-world 127.0.0.1/mytest/hello-world:v1
```

命令运行结果如图 6-14 所示。

```
[root@docker ~]# docker tag hello-world 127.0.0.1/mytest/hello-world:v1
[root@docker ~]# docker images
REPOSITORY                      TAG       IMAGE ID       CREATED         SIZE
goharbor/harbor-exporter        v2.9.1    37bfd4fa26bc   4 weeks ago     105MB
goharbor/redis-photon           v2.9.1    67827413c0fd   4 weeks ago     209MB
goharbor/trivy-adapter-photon   v2.9.1    a02695b8f8ea   4 weeks ago     469MB
goharbor/harbor-registryctl     v2.9.1    a076218bb631   4 weeks ago     148MB
goharbor/registry-photon        v2.9.1    2f01ea8b1853   4 weeks ago     82.7MB
goharbor/nginx-photon           v2.9.1    5200203dd7ef   4 weeks ago     153MB
goharbor/harbor-log             v2.9.1    ac1cdcc94a5f   4 weeks ago     162MB
goharbor/harbor-jobservice      v2.9.1    d9ff6fc98cc8   4 weeks ago     139MB
goharbor/harbor-core            v2.9.1    0a3a7953409c   4 weeks ago     166MB
goharbor/harbor-portal          v2.9.1    345284db8ca1   4 weeks ago     161MB
goharbor/harbor-db              v2.9.1    69606d285be1   4 weeks ago     358MB
goharbor/prepare                v2.9.1    adb2d804c458   4 weeks ago     253MB
nginx                           latest    c20060033e06   4 weeks ago     187MB
httpd                           new       542540f88e3f   7 weeks ago     505MB
tomcat                          new       fd0b1833593e   8 weeks ago     615MB
nginx                           new       b12a1a921e1f   20 months ago   471MB
sshd                            new       a7742ca6b111   20 months ago   365MB
127.0.0.1/mytest/hello-world    v1        feb5d9fea6a5   2 years ago     13.3kB
hello-world                     latest    feb5d9fea6a5   2 years ago     13.3kB
```

图 6-14　将本地已有镜像 hello-world 打上标签

第七步，上传镜像到 Harbor 仓库，操作命令如下：

```
[root@docker ~]# docker push 127.0.0.1/mytest/hello-world:v1
```

命令运行结果如图 6-15 所示。

```
[root@docker ~]# docker push 127.0.0.1/mytest/hello-world:v1
The push refers to repository [127.0.0.1/mytest/hello-world]
e07ee1baac5f: Pushed
v1: digest: sha256:f54a58bc1aac5ea1a25d796ae155dc228b3f0e11d046ae276b39c4bf2f13d8c4 size: 525
```

图 6-15　上传镜像到 Harbor 仓库

第八步，在 Harbor 界面的 mytest 项目中可以看到该镜像及其相关信息，如图 6-16 所示。

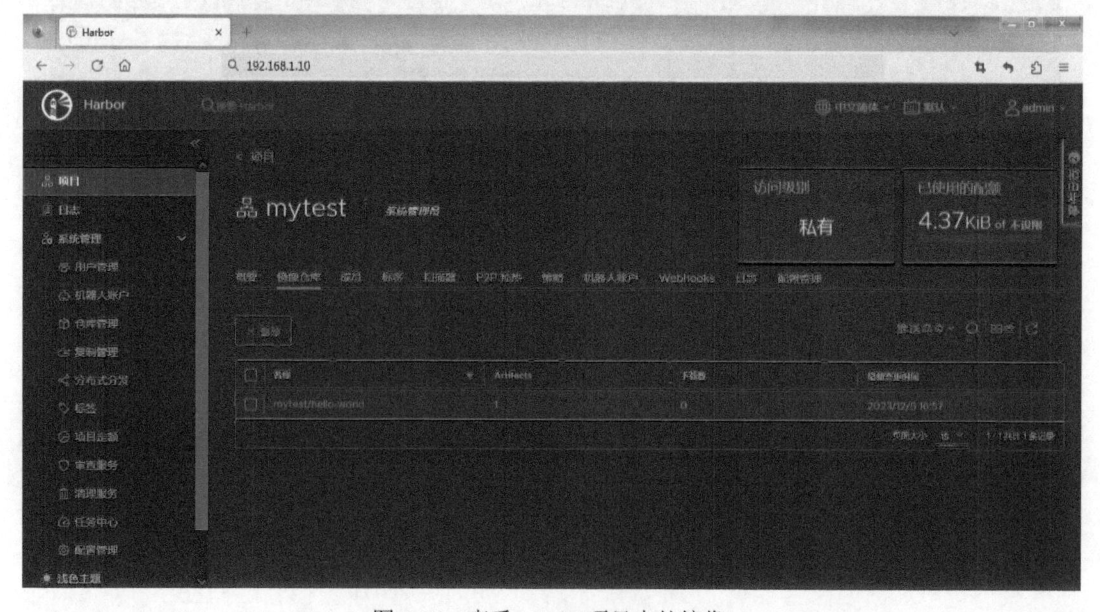

图 6-16　查看 mytest 项目中的镜像

第九步，开启 Client 客户机，配置其 Docker 服务，操作命令如下：

```
[root@client ~]# vim /usr/lib/systemd/system/docker.service
```

在文本中修改的代码如下：

```
ExecStart=/usr/bin/dockerd --insecure-registry 192.168.1.10
```

修改之后的效果如图 6-17 所示。

```
[Unit]
Description=Docker Application Container Engine
Documentation=https://docs.docker.com
After=network-online.target docker.socket firewalld.service containerd.service
Wants=network-online.target
Requires=docker.socket containerd.service

[Service]
Type=notify
# the default is not to use systemd for cgroups because the delegate issues still
# exists and systemd currently does not support the cgroup feature set required
# for containers run by docker
# ExecStart=/usr/bin/dockerd -H fd:// --containerd=/run/containerd/containerd.sock
ExecStart=/usr/bin/dockerd --insecure-registry 192.168.1.10
ExecReload=/bin/kill -s HUP $MAINPID
TimeoutSec=0
RestartSec=2
Restart=always

# Note that StartLimit* options were moved from "Service" to "Unit" in systemd 229.
# Both the old, and new location are accepted by systemd 229 and up, so using the old location
# to make them work for either version of systemd.
StartLimitBurst=3
```

图 6-17　在 client 客户机配置 docker 服务

第十步，重新加载服务，重启 Docker，操作命令如下：

```
[root@client ~]# systemctl daemon-reload
[root@client ~]# systemctl restart docker
```

第十一步，登录 Harbor，操作命令如下：

```
[root@client ~]# docker login -u admin -p Harbor12345 http://192.168.1.10
```

命令运行结果如图 6-18 所示。

```
[root@client ~]# systemctl daemon-reload
[root@client ~]# systemctl restart docker
[root@client ~]# docker login -u admin -p Harbor12345 http://192.168.1.10
WARNING! Using --password via the CLI is insecure. Use --password-stdin.
WARNING! Your password will be stored unencrypted in /root/.docker/config.json.
Configure a credential helper to remove this warning. See
https://docs.docker.com/engine/reference/commandline/login/#credentials-store

Login Succeeded
```

图 6-18　登录 Harbor

第十二步，拉取镜像 hello-world，并给该镜像打上标签，操作命令如下：

```
[root@client ~]# docker pull hello-world
[root@client ~]# docker tag hello-world 192.168.1.10/mytest/hello-world:v2
```

命令运行结果如图 6-19 所示。

```
[root@client ~]# docker pull hello-world
Using default tag: latest
latest: Pulling from library/hello-world
719385e32844: Pull complete
Digest: sha256:c79d06dfdfd3d3eb04cafd0dc2bacab0992ebc243e083cabe208bac4dd7759e0
Status: Downloaded newer image for hello-world:latest
docker.io/library/hello-world:latest
[root@client ~]# docker tag hello-world 192.168.1.10/mytest/hello-world:v2
[root@client ~]# docker images
REPOSITORY                        TAG      IMAGE ID       CREATED          SIZE
mysshd                            v1       140760ec6568   4 months ago     341MB
192.168.1.10/mytest/hello-world   v2       9c7a54a9a43c   7 months ago     13.3kB
hello-world                       latest   9c7a54a9a43c   7 months ago     13.3kB
sshd                              v2       958d5180baa6   16 months ago    341MB
sshd                              new      a7742ca6b111   20 months ago    365MB
mysql                             5.7      f26e21ddd20d   20 months ago    450MB
```

图 6-19　拉取镜像并打标签

第十三步，上传该镜像到 Harbor 的 mytest 项目中，操作命令如下：

```
[root@docker ~]# docker push 192.168.1.10/mytest/hello-world:v2
```

命令运行结果如图 6-20 所示。

```
[root@client ~]# docker push 192.168.1.10/mytest/hello-world:v2
The push refers to repository [192.168.1.10/mytest/hello-world]
01bb4fce3eb1: Layer already exists
v2: digest: sha256:7e9b6e7ba2842c91cf49f3e214d04a7a496f8214356f41d81a6e6dcad11f11e3 size: 525
```

图 6-20　上传镜像到 Harbor 仓库

第十四步，查看 Harbor 管理界面的 mytest 项目，其中已有两个项目，如图 6-21 所示。

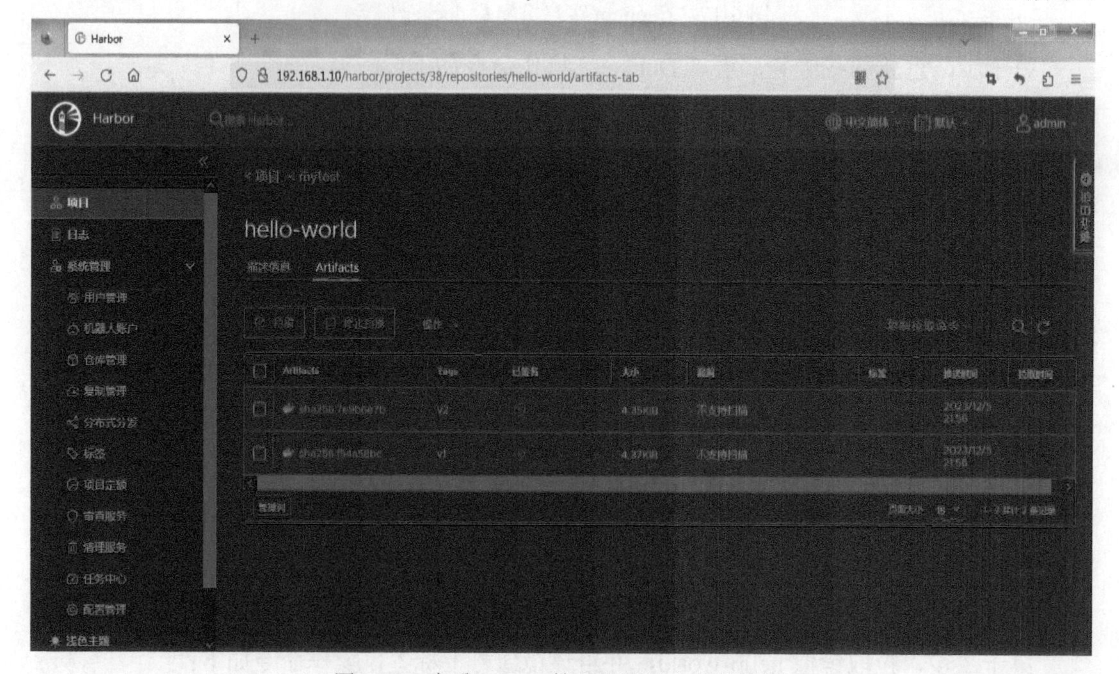

图 6-21　查看 Harbor 管理界面已上传的镜像

第十五步，创建用户。用鼠标单击 Harbor 管理界面左侧“系统管理”下“用户管理”，点击“+ 创建用户”，在对话框中输入用户名 test-user01，邮箱 test-user0 @yjsd.cn，全名云计算新星，密码 Admin123，注释为管理员，如图 6-22 所示。

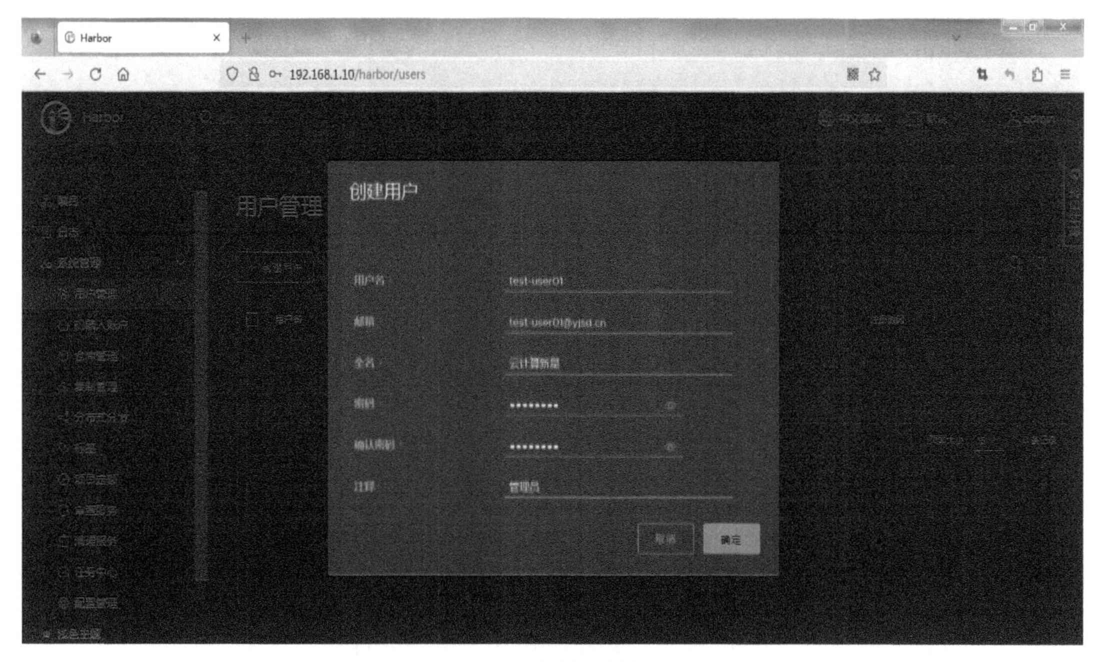

图 6-22　填写创建用户信息

第十六步，单击确定按钮，创建用户成功，用同样的方法创建用户 test-user02，注释为程序员，如图 6-23 所示。

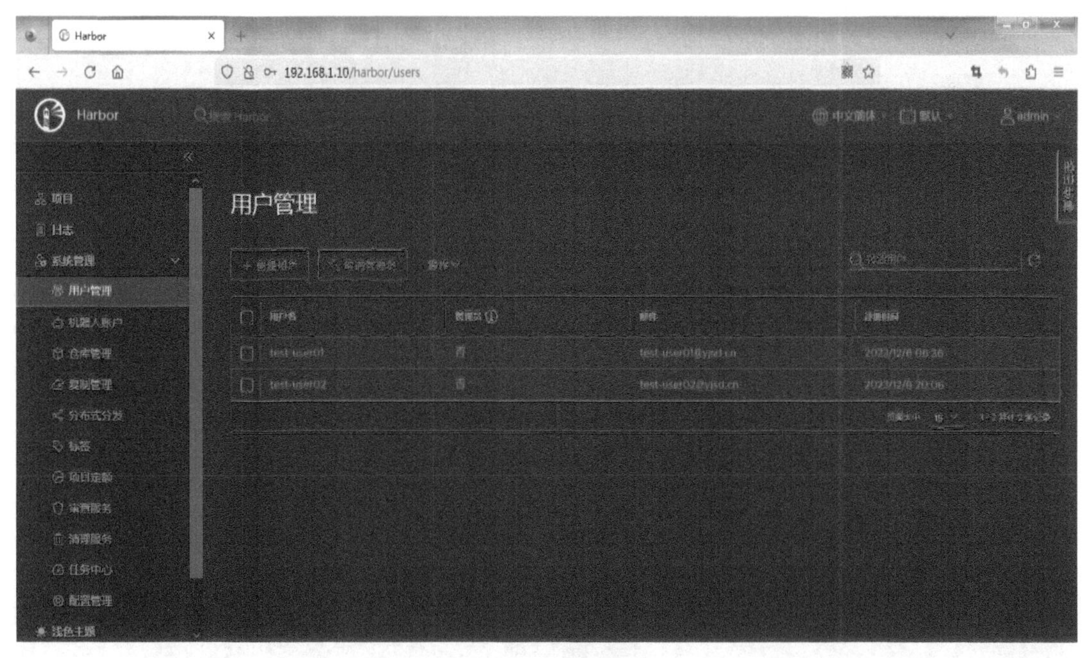

图 6-23　创建用户成功

第十七步，配置用户权限。选择 test-user01 用户，单击“设置为管理员”，将该用户设置为管理员角色，如图 6-24 所示。

图 6-24　配置 test-user01 用户为管理员角色

第十八步，添加项目成员。单击 Harbor 管理界面的“项目”和“mytest”选项，在“mytest”界面下点击“成员”“+ 用户”，填写用户名称 test-user02，并为其分配角色为开发者，如图 6-25 所示。

图 6-25　新建项目成员

第十九步，单击确定按钮，成功添加项目成员，如图 6-26 所示。

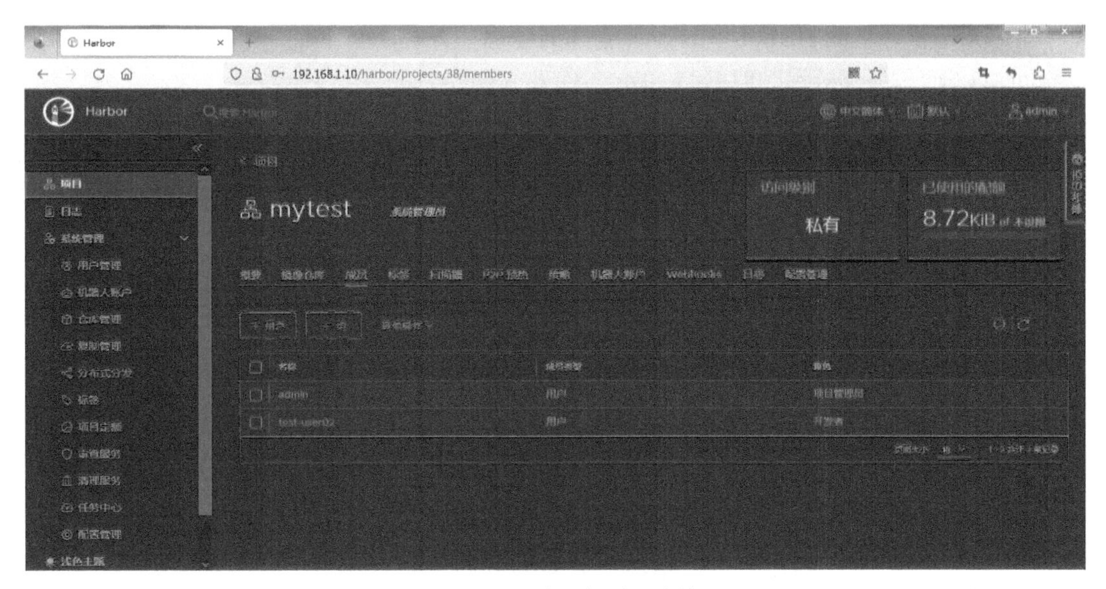

图 6-26　成功添加项目成员

第二十步，在 Client 客户机上删除上述打标签的本地镜像 hello-world:v2，操作命令如下：

[root@client ~]# docker rmi 192.168.1.10/mytest/hello-world:v2

命令运行结果如图 6-27 所示。

```
[root@client ~]# docker images
REPOSITORY                          TAG       IMAGE ID       CREATED         SIZE
mysshd                              v1        140760ec6568   4 months ago    341MB
hello-world                         latest    9c7a54a9a43c   7 months ago    13.3kB
192.168.1.10/mytest/hello-world     v2        9c7a54a9a43c   7 months ago    13.3kB
sshd                                v2        958d5180baa6   16 months ago   341MB
sshd                                new       a7742ca6b111   20 months ago   365MB
mysql                               5.7       f26e21ddd20d   20 months ago   450MB
centos                              7         eeb6ee3f44bd   2 years ago     204MB
centos/httpd                        latest    2cc07fbb5000   4 years ago     258MB
[root@client ~]# docker rmi 192.168.1.10/mytest/hello-world:v2
Untagged: 192.168.1.10/mytest/hello-world:v2
Untagged: 192.168.1.10/mytest/hello-world@sha256:7e9b6e7ba2842c91cf49f3e214d04a7a496f8214356f41d81a6e6dcad11f11e3
[root@client ~]# docker images
REPOSITORY     TAG       IMAGE ID       CREATED         SIZE
mysshd         v1        140760ec6568   4 months ago    341MB
hello-world    latest    9c7a54a9a43c   7 months ago    13.3kB
sshd           v2        958d5180baa6   16 months ago   341MB
sshd           new       a7742ca6b111   20 months ago   365MB
mysql          5.7       f26e21ddd20d   20 months ago   450MB
centos         7         eeb6ee3f44bd   2 years ago     204MB
centos/httpd   latest    2cc07fbb5000   4 years ago     258MB
```

图 6-27　删除上述打标签的本地镜像

第二十一步，退出当前用户，使用 test-user01 用户登录，命令如下：

[root@client ~]# docker logout 192.168.1.10

[root@client ~]# docker login 192.168.1.10

命令运行结果如图 6-28 所示。

```
[root@client ~]# docker logout 192.168.1.10
Removing login credentials for 192.168.1.10
[root@client ~]# docker login 192.168.1.10
Username: test-user01
Password:
WARNING! Your password will be stored unencrypted in /root/.docker/config.json.
Configure a credential helper to remove this warning. See
https://docs.docker.com/engine/reference/commandline/login/#credentials-store

Login Succeeded
```

图 6-28　退出当前用户，使用 test-user01 用户登录

第二十二步，下载 Harbor 中 mytest 仓库的镜像并查看，操作命令如下：

```
[root@client ~]# docker pull 192.168.1.10/mytest/hello-world:v1
[root@client ~]# docker images
```

命令运行结果如图 6-29 所示。

```
[root@client ~]# docker pull 192.168.1.10/mytest/hello-world:v1
v1: Pulling from mytest/hello-world
2db29710123e: Pull complete
Digest: sha256:f54a58bc1aac5ea1a25d796ae155dc228b3f0e11d046ae276b39c4bf2f13d8c4
Status: Downloaded newer image for 192.168.1.10/mytest/hello-world:v1
192.168.1.10/mytest/hello-world:v1
[root@client ~]# docker images
REPOSITORY                        TAG       IMAGE ID       CREATED         SIZE
mysshd                            v1        140760ec6568   4 months ago    341MB
hello-world                       latest    9c7a54a9a43c   7 months ago    13.3kB
sshd                              v2        958d5180baa6   16 months ago   341MB
sshd                              new       a7742ca6b111   20 months ago   365MB
mysql                             5.7       f26e21ddd20d   20 months ago   450MB
192.168.1.10/mytest/hello-world   v1        feb5d9fea6a5   2 years ago     13.3kB
```

图 6-29　下载 Harbor 中 mytest 仓库的镜像并查看

第二十三步，在 Harbor 管理界面单击右上角“事件日志”，查看用户相关操作，如图 6-30 所示。

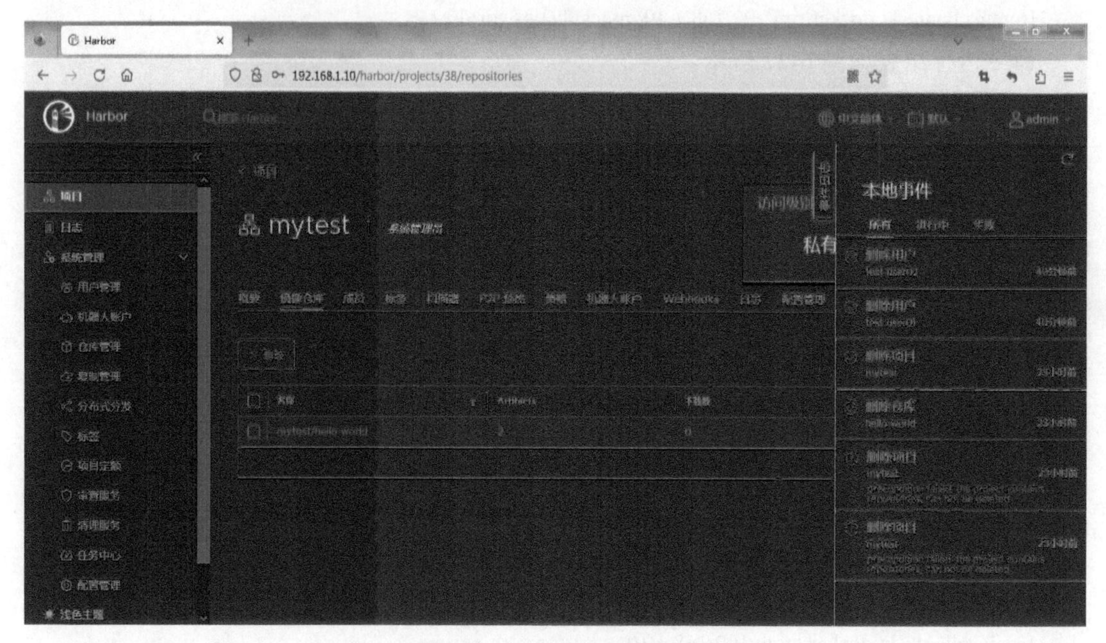

图 6-30　查看用户相关操作

第二十四步，在 Harbor 服务器的 harbor 目录中，可以使用 docker-compose 管理 Harbor，停止 / 启动 / 重启操作的命令如下：

```
[root@docker ~]# cd /usr/local/harbor
[root@docker harbor]# docker-compose stop
[root@docker harbor]# docker-compose start
[root@docker harbor]# docker-compose restart
```

双创视角

腾讯云在微信支付中的应用

微信支付商户服务平台为数千万商家提供下载账单、查询账单并进行统计分析服务，起初该平台将数据存储在一个开源的 MySQL 数据库中，但是随着京东等大型商户加入该平台，交易量的增加很容易耗尽独立数据库的有限存储空间，给微信支付造成了严重的容量和性能瓶颈。

腾讯云数据库团队经过多年的努力，基于开源的 PostgreSQL 数据库为企业开发了一站式数据库解决方案——TDSQL for PostgreSQL，它是一个能够进行混合事务分析处理 (Hybrid Transaction/Analytical Processing，HTAP) 的分布式数据库管理系统。

TDSQL for PostgreSQL 帮助微信支付升级到一个数据密集型应用程序的水平，支持越来越多的场景，保证业务的稳定性和连续性；在并行编写性能和吞吐量方面具有明显的优势，大大减少了数据的写入时间，并满足了构建实时报告的要求；在读取报告数据方面，将模糊搜索商家结果所需要的时间从大约需要 17 s 提升到不到 50 ms。

到目前为止，微信支付已经在 TDSQL for PostgreSQL 数据库中存储了超过 400 TB 的数据，每秒处理超过 24 万个请求，其中 99.6% 的处理时间控制在 10 ms 的范围内，完全满足了具有挑战性的性能和稳定性要求。

项 目 小 结

本项目介绍了公有仓库和 Harbor 私有仓库的相关知识，完成了部署和管理 Harbor 私有仓库的操作任务，让读者学会 Harbor 私有仓库的使用方法。

习 题 测 试

一、单选题

1. Docker 官方给用户提供了一个 Docker 仓库，它就像手机里的 (　　) 一样，里面存放着各种各样已经打包好的应用——Docker 镜像。

A. 图库　　B. 备忘录
C. 应用市场　　D. 实用工具

2. (　　) 用于镜像的存储，是 Docker 镜像分发和部署的关键。

A. Docker 仓库　　B. 本地镜像库
C. Docker Compose　　D. Harbor

3. Docker 运行中使用的默认仓库就是 (　　)。

A. 容器卷　　B. 本地镜像库
C. Harbor 公开仓库　　D. Docker Hub 公共仓库

二、多选题

1. Docker 的三大基础组件是（　　）。

A. 镜像　　B. 容器

C. 仓库　　D. 脚本

2. 以下（　　）是 Harbor 的特性。

A. 图形化 UI　　B. 审计管理

C. 基于角色控制　　D. 镜像删除和空间回收

3. Harbor 的每个组建均以（　　）形式构建，因此使用（　　）对其进行部署。

A. Docker 镜像　　B. Docker 容器

C. Docker Compose　　D. Docker Swarm

三、简答题

1. 简述 Docker Hub 公有仓库的作用。

2. 简述 Harbor 各组件的作用。

项目七
部署 Docker 安全

学习目标

(1) 了解 Docker 安全。
(2) 了解 Kubernetes 的体系结构。
(3) 理解 Kubernetes 的核心概念。
(3) 掌握部署 Kubernetes 系统环境的方法。
(4) 掌握搭建 Kubernetes 集群的方法。

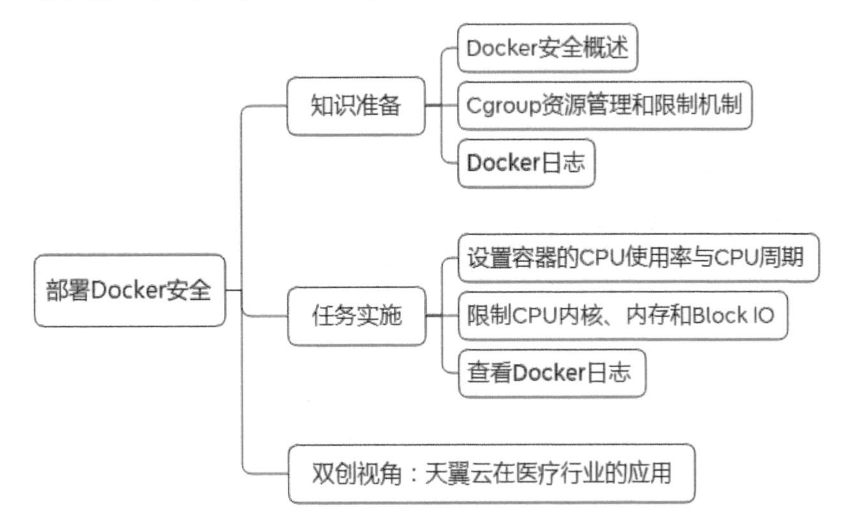

7.1 知识准备

7.1.1 Docker 安全概述

1. Docker 容器与虚拟机的安全性比较

Docker 容器与宿主机共享操作系统，容器中的应用有可能导致虚拟机崩溃，而虚拟机崩溃一般不会影响宿主机的运行，如图 7-1 所示。Docker Engine 引擎是运行和管理容器的核心软件，通常会简单地指 Docker，用以创建和运行容器。Docker 使用 cgroups、namespaces 和 SELinux/AppArmor 安全策略等多种技术来实现容器隔离，让应用能够在容器内运行而不影响宿主机和其他容器。Hypervisor 处于硬件系统之上，它将 CPU、内存、网卡等转换为虚拟资源按需分配给每个虚拟机，每个虚拟机都配置自己的操作系统，想通过虚拟机攻击宿主机或其他虚拟机，必须经过 Hypervisor 层，而这个难度是相当大的，所以与 Docker 容器相比，虚拟机的安全性比较高。

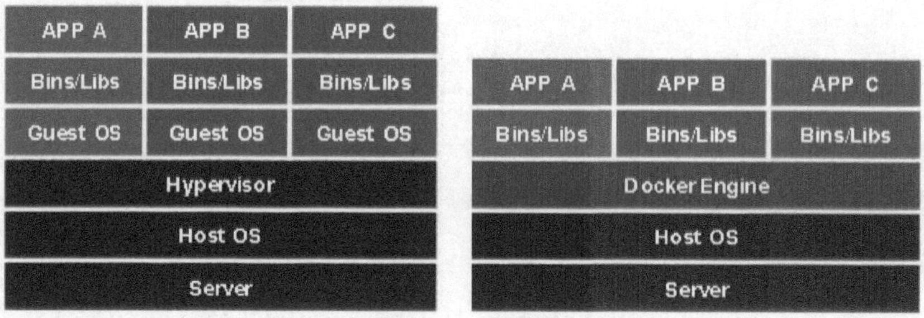

图 7-1 容器与虚拟机

2. 影响 Docker 安全的因素

以下是常见的导致 Docker 安全性问题的一些因素：

(1) 版本漏洞：虽然 Docker 版本在不断升级，但 Docker 引擎和相关组件仍然可能存在漏洞，攻击者可能利用这些漏洞对系统进行攻击。

(2) 未经授权的访问：未经授权的用户或者容器可能会访问其他容器或者宿主机上的敏感信息。

(3) 不安全的镜像：通过公开的镜像网站，使用未经验证或者来源不明的镜像中可能包含恶意代码，攻击者可以通过恶意镜像从而进行攻击。

(4) 网络通信：容器之间的网络通信可能不安全，攻击者可以通过监听、拦截或者篡改网络流量来进行攻击。

(5) 资源耗尽：攻击者控制的恶意容器可能会消耗过多的系统资源，导致拒绝服务或者其他系统性能问题。

(6) 缺乏监控和日志：如果缺乏对容器活动的监控和日志记录，可能导致无法追踪或安全分析事件。

除此之外，还有不完善的配置、存储泄漏、容器逃逸、不适当的权限管理等，也会导致 Docker 的安全性问题。

7.1.2　Cgroup 资源管理和限制机制

Cgroup(Control Groups，控制族群) 是 Linux 内核提供的一种资源管理和限制机制，它能提供统一的接口来管理 CPU、内存、磁盘 I/O、网络等资源，可以对进程进行分组并对分组内的进程进行资源限制、优先级调整等操作，从而更好地控制系统中各个进程的资源使用情况，实现资源隔离和共享，避免因某些进程占用资源过多而导致系统负载过高的问题。

1. Cgroup 的相关概念

任务 (task)：系统的一个进程。一个任务可以是多个不同层级 Cgroup 的成员。

控制组 (Cgroup)：按照一定规则划分的一组进程。一个进程可以加入到某个控制组，Cgroup 以控制组为单位进行资源分配和限制。

层级树 (hierarchy)：按照树状结构排列组成的一系列控制组，即一棵控制树。控制组中的子节点是父节点的孩子，继承父节点的属性。例如，系统中定义了控制组 C1，限制 CPU 使用 2 核，然后再定义控制组 C2，既限制 CPU 使用 2 核又限制内存使用 4G，则控制组 C1 和 C2 可以组成父子关系，无须重复定义 CPU 限制，直接继承 C1 即可。

子系统 (subsystem)：用以实现对某类资源的管理和控制，一个子系统就是一个资源控制器 (controllers)。例如，CPU 子系统是限制 CPU 使用时间的控制器，内存子系统是限制内存使用量的控制器。每个控制组可以关联一个或多个子系统，子系统必须附加到一个层级上才能起作用，层级上的所有控制组都受到该子系统的控制。

份额：默认一个 Docker 容器的 CPU 份额是 1024，当同时运行两个容器 container1 和 container2 时，如果其 CPU 份额分别为 512 和 1024(即份额比为 1∶2)，则 container2 获得 CPU 的使用率比 container1 大一倍。但如果 container2 一直处于空闲，则 container1 可以获得更多的 CPU 使用率。当主机上只运行一个容器时，即使其 CPU 份额只有 30，它也可以独占整个主机的 CPU 资源。Cgroup 只在容器占用资源紧缺时才起作用，Cgroup 资源配置取决于多个运行容器的 CPU 分配情况和容器中进程的运行情况。

2. Cgroup 子系统

在 /sys/fs/cgroup/ 目录下，可以看到 Cgroup 子系统，如图 7-2 所示。

```
[root@docker ~]# ll /sys/fs/cgroup
total 0
drwxr-xr-x 5 root root  0 May  8 21:42 blkio
lrwxrwxrwx 1 root root 11 May  8 21:42 cpu -> cpu,cpuacct
lrwxrwxrwx 1 root root 11 May  8 21:42 cpuacct -> cpu,cpuacct
drwxr-xr-x 5 root root  0 May  8 21:42 cpu,cpuacct
drwxr-xr-x 3 root root  0 May  8 21:42 cpuset
drwxr-xr-x 5 root root  0 May  8 21:42 devices
drwxr-xr-x 3 root root  0 May  8 21:42 freezer
drwxr-xr-x 3 root root  0 May  8 21:42 hugetlb
drwxr-xr-x 5 root root  0 May  8 21:42 memory
lrwxrwxrwx 1 root root 16 May  8 21:42 net_cls -> net_cls,net_prio
drwxr-xr-x 3 root root  0 May  8 21:42 net_cls,net_prio
lrwxrwxrwx 1 root root 16 May  8 21:42 net_prio -> net_cls,net_prio
drwxr-xr-x 3 root root  0 May  8 21:42 perf_event
drwxr-xr-x 5 root root  0 May  8 21:42 pids
drwxr-xr-x 5 root root  0 May  8 21:42 systemd
[root@docker ~]#
```

图 7-2　虚拟机与容器

Cgroup 子系统是一组资源控制模块，它们的作用分别如下：

blkio：为块设备设置 I/O 限制。

cpu：使用调度程序设置进程的 CPU 占用时间。

cpuacct：自动生成 Cgroup 中任务所使用 CPU 资源情况报告。

cpuset：为 Cgroup 中的任务分配独立的 CPU(多核系统) 和内存节点。

devices：开启或关闭 Cgroup 中任务对设备的访问。

freezer：挂起或恢复 Cgroup 中的任务。

hugetlb：限制使用的内存页数量。

memory：设置 Cgroup 中任务对内存占用的限定。

net_cls：使用等级识别符 (classid) 标记网络数据包，将 Cgroup 中进程产生的网络包分类，让 Linux 流量控制器 tc 可以根据分类区分数据包并做网络限制。

net_prio：设置 Cgroup 中进程产生网络流量的优先级。

perf_event：使用 perf 工具监控 Cgroup。

pids：限制任务数量。

systemd：提供 Cgroup 使用和管理接口。

7.1.3　Docker 日志

在实际的生产环境中，高效监控和日志管理对系统的持续稳定运行以及故障排查至关重要。

1. Docker 日志的主要作用

性能监控：日志记录了容器运行时的 CPU 使用率、内存占用、网络流量等信息，通过分析这些数据，可以对容器的性能进行监控，及时发现性能瓶颈，优化应用性能。

问题排查：当应用出现问题时，查看 Docker 容器的日志，分析日志中的错误信息，有助于快速定位问题的根源，从而采取相应的解决措施。

安全审计：日志记录了容器的启动、停止、访问控制等信息，据此可以检测应用中潜在的安全风险，防患于未然。

故障恢复：当系统发生故障时，通过查看和分析日志，可以了解故障发生前后的容器状态，有助于快速恢复系统的正常运行状态。

优化改进：通过对日志进行分析，可以发现应用的运行情况和潜在问题，为应用的优化和改进提供依据。

2. Docker 日志的分类

Docker 日志分为 Docker 引擎日志和容器日志两类。

Docker 引擎日志是指 Docker 运行时所产生的日志，一般存储在 /var/log/docker.log 文件中，或使用 journalctl -u docker 命令查看，具体位置因系统而异。

Docker 容器日志是指容器在运行过程所产生的日志，它记录了各种输出、警告或错误等信息，以便用户对容器进行监控、性能分析和故障排查。每个容器的日志一般存储在 /var/lib/docker/containers/<container-id>/container-json.log 文件中。

3. Docker 日志的第三方管理工具

在实际应用中，有时需要更高级的日志管理工具对大规模集群的日志进行处理，如集中化管理、实时搜索、统计分析等。Docker 日志常见的第三方管理工具包括 Sysdig、Splunk、ELK 等。

Sysdig 是一个开源的系统追踪工具，能够实时捕获和分析 Linux 系统和容器内部信息，并生成详细的事件日志。它还提供了一个基于 Web 的用户界面，对容器、进程、文件系统、网络等进行实时的性能监控。

Splunk 是一款商业化的日志管理和分析平台，提供实时监控、搜索和可视化等功能，支持在 Docker 容器中监控和优化应用程序的性能。

ELK 是一个流行的日志管理和分析平台，由 Elasticsearch、Logstash 和 Kibana 三个开源工具组成，可以为用户提供强大的 Docker 日志收集、存储、搜索和可视化等功能。

这些工具各有特点，应用于不同的生产场景，可以根据用户的具体需求选择合适的工具进行 Docker 容器的日志管理。

7.2 任务实施

7.2.1 设置容器的 CPU 使用率与 CPU 周期

设置容器的 CPU 使用率与 CPU 周期

1. 任务目标

(1) 掌握设置容器占用 CPU 使用率的方法。

(2) 掌握设置 CPU 周期的方法。

2. 任务内容

(1) 创建 Stress 工具镜像。

(2) 设置容器占用 CPU 资源的份额。

(3) 设置 CPU 周期。

3. 完成任务所需的设备和软件

(1) 一台安装 Windows 10 操作系统的计算机。

(2) VMware Workstation、Docker。

(3) 远程管理工具 MobaXterm。

4. 任务实施步骤

第一步，使用 Dockerfile 创建 Stress 工具镜像，操作命令如下：

```
[root@docker ~]# docker pull centos:7
[root@docker ~]# mkdir stress
[root@docker ~]# cd stress/
[root@docker stress]# vim Dockerfile
```

Dockerfile 代码如下：

```
FROM centos:7
MAINTAINER zsk "zsk@kgc.com"
RUN yum -y install wget && \
wget -O /etc/yum.repos.d/epel.repo http://mirrors.aliyun.com/repo/epel-7.repo \
&& yum -y install stress
[root@docker stress]# docker build -t centos:stress .
[root@docker stress]# docker images
```

命令运行结果如图 7-3 所示。

```
[root@docker stress]# docker images
REPOSITORY                     TAG        IMAGE ID        CREATED           SIZE
centos                         stress     43afa3c3b8d2    10 minutes ago    497MB
```

图 7-3　创建 Stress 工具镜像

第二步，启动容器 container1，查看 CPU 的使用百分比。此处模拟系统负载较高时的场景，开启了 8 个 stress 进程，从而让 Cgroup 生效，操作命令如下：

```
[root@docker ~]# docker run -tid --name container1 --cpu-shares 521 centos:stress stress -c 8
```

“--cpu-shares”表示同时运行多个容器时，用于分配容器所占用的 CPU 份额。

```
[root@docker ~]# docker ps -a
[root@docker ~]# docker exec -it container1 bash
[root@53654d0c7f97 /]# top
```

命令运行结果如图 7-4 所示。

```
[root@docker ~]# docker run -tid --name container1 --cpu-shares 521 centos:stress stress -c 8
53654d0c7f972f3db856a833dc3db2c5c067e867010da64f7cbe7d53913aad75
[root@docker ~]# docker ps -a
CONTAINER ID   IMAGE           COMMAND         CREATED          STATUS          PORTS     NAMES
53654d0c7f97   centos:stress   "stress -c 8"   22 seconds ago   Up 20 seconds             container1
[root@docker ~]# docker exec -it container1 bash
[root@53654d0c7f97 /]# top
top - 06:07:20 up  4:50,  0 users,  load average: 8.77, 11.53, 14.46
Tasks:  11 total,   9 running,   2 sleeping,   0 stopped,   0 zombie
%Cpu(s):100.0 us,  0.0 sy,  0.0 ni,  0.0 id,  0.0 wa,  0.0 hi,  0.0 si,  0.0 st
KiB Mem :  1863252 total,  1101264 free,   279344 used,   482644 buff/cache
KiB Swap:  2097148 total,  2097148 free,        0 used.  1424680 avail Mem

  PID USER      PR  NI    VIRT    RES    SHR S  %CPU %MEM     TIME+ COMMAND
   13 root      20   0    7312    100      0 R  25.6  0.0   0:14.71 stress
    9 root      20   0    7312    100      0 R  25.2  0.0   0:14.83 stress
    6 root      20   0    7312    100      0 R  24.9  0.0   0:14.77 stress
    7 root      20   0    7312    100      0 R  24.9  0.0   0:14.68 stress
    8 root      20   0    7312    100      0 R  24.9  0.0   0:14.71 stress
   11 root      20   0    7312    100      0 R  24.9  0.0   0:14.78 stress
   10 root      20   0    7312    100      0 R  24.6  0.0   0:14.85 stress
   12 root      20   0    7312    100      0 R  24.3  0.0   0:14.75 stress
    1 root      20   0    7312    632    532 S   0.0  0.0   0:00.04 stress
   14 root      20   0   11804   1888   1488 S   0.0  0.1   0:00.04 bash
   28 root      20   0   56184   1992   1440 R   0.0  0.1   0:00.00 top
```

图 7-4　查看 container1 资源占用情况

(1) 系统显示信息说明，见表 7-1～表 7-5 所示。

表 7-1　top：系统运行时间和平均负载

显示信息	说　明
- 02:45:39	当前的时间
up 4:50	已运行的时间
0 users	当前登录的用户数
load average: 8.77, 11.53, 14.46	系统负载的平均值：即过去 1 分钟、5 分钟、15 分钟内的均值

表 7-2　Tasks：当前运行的进程情况

显示信息	说　明
11 total	进程的总数
9 running	正在运行的进程数
2 sleeping	睡眠进程数
0 stopped	停止进程数
0 zombie	僵尸进程数

表 7-3　%Cpu(s)：CPU 的使用率

显示信息	说　明
100.0 us	用户进程的 CPU 使用率
0.0 sy	系统进程的 CPU 使用率
0.0 ni	用户改变的优先级进程的 CPU 使用率
0.0 id	空闲进程的 CPU 使用率
0.0 wa	等待 I/O 的 CPU 使用率
0.0 hi	硬件中断的 CPU 使用率
0.0 si	软件中断的 CPU 使用率
0.0 st	抢断 CPU 的使用率

表 7-4　KiB Mem：物理内存使用情况

显示信息	说　明
1863252 total	物理总内存 (默认单位为 KB)
1101356 free	空闲物理内存
279344 used	已使用物理内存
481556 buff/cache	用作缓存的内存

表 7-5　KiB Swap：交换区的使用情况

显示信息	说　明
2097148 total	交换区的总量 (默认单位为 KB)
2097148 free	空闲的交换区
0 used	已使用的交换区
1424680 avail Mem	缓冲交换区

(2) 进程表格信息说明，见表 7-6 所示。

表 7-6　进程表格信息说明

显示信息	说　明
PID	进程 ID
USER	用户名
PR	进程优先级
NI	用户进程 nice 值，负值表示高优先级，正值表示低优先级
VIRT	进程占用虚拟内存的大小，单位为 KB，VIRT = SWAP + RES
RES	常驻内存的大小，单位为 KB，RES = CODE(可执行代码占用的物理内存) + DATA
SHR	共享内存的大小，单位为 KB
S	进程状态，R 表示运行，S 表示睡眠，D 表示不可中断的睡眠状态，T 表示跟踪或停止，Z 表示僵尸进程
%CPU	CPU 的使用率
%MEM	内存使用率
TIME+	累计使用 CPU 的时间，单位为 1/100 s
COMMAND	运行命令

第三步，启动容器 container2，查看 CPU 的使用百分比，操作命令如下：

```
[root@docker ~]# docker run -tid --name container2 --cpu-shares 1024 centos:stress stress -c 8
[root@docker ~]# docker ps -a
[root@docker ~]# docker exec -it container2 bash
[root@32894aeeda10 /]# top
```

命令运行结果如图 7-5 所示。

```
[root@docker ~]# docker run -tid --name container2 --cpu-shares 1024 centos:stress stress -c 8
32894aeeda10233e91e05b1d57bec99a04c5bf25917c01346230d0c6b6d750da
[root@docker ~]# docker ps -a
CONTAINER ID   IMAGE           COMMAND         CREATED          STATUS         PORTS     NAMES
32894aeeda10   centos:stress   "stress -c 8"   8 seconds ago    Up 6 seconds             container2
53654d0c7f97   centos:stress   "stress -c 8"   3 minutes ago    Up 2 minutes             container1
[root@docker ~]# docker exec -it container2 bash
[root@32894aeeda10 /]# top
top - 06:09:49 up  4:53,  0 users,  load average: 11.70, 11.06, 13.80
Tasks:  11 total,   9 running,   2 sleeping,   0 stopped,   0 zombie
%Cpu(s):100.0 us,  0.0 sy,  0.0 ni,  0.0 id,  0.0 wa,  0.0 hi,  0.0 si,  0.0 st
KiB Mem :  1863252 total,  1087912 free,   292384 used,   482956 buff/cache
KiB Swap:  2097148 total,  2097148 free,        0 used.  1411484 avail Mem

  PID USER      PR  NI    VIRT    RES    SHR S  %CPU %MEM     TIME+ COMMAND
   13 root      20   0    7312    100      0 R  16.9  0.0   0:06.05 stress
    7 root      20   0    7312    100      0 R  16.6  0.0   0:06.05 stress
    8 root      20   0    7312    100      0 R  16.6  0.0   0:06.05 stress
    9 root      20   0    7312    100      0 R  16.6  0.0   0:06.03 stress
   11 root      20   0    7312    100      0 R  16.6  0.0   0:06.14 stress
   12 root      20   0    7312    100      0 R  16.6  0.0   0:06.11 stress
   10 root      20   0    7312    100      0 R  16.3  0.0   0:06.03 stress
   14 root      20   0    7312    100      0 R  16.3  0.0   0:06.05 stress
    1 root      20   0    7312    632    532 S   0.0  0.0   0:00.05 stress
   15 root      20   0   11804   1884   1488 S   0.0  0.1   0:00.04 bash
   29 root      20   0   56184   1988   1440 R   0.0  0.1   0:00.00 top
```

图 7-5　查看 container2 资源占用情况

第四步，重新进入容器 container1，对比两个容器的 CPU 使用百分比，操作命令如下：

```
[root@docker ~]# docker exec -it container1 bash
```

[root@53654d0c7f97 /]# top

命令运行结果如图 7-6 所示。可见两容器 container1 和 container2 的 CPU 使用率比值为 1∶2。

```
[root@docker ~]# docker exec -it container1 bash
[root@53654d0c7f97 /]# top
top - 06:12:27 up  4:56,  0 users,  load average: 15.93, 13.21, 14.19
Tasks:  11 total,   9 running,   2 sleeping,   0 stopped,   0 zombie
%Cpu(s):100.0 us,  0.0 sy,  0.0 ni,  0.0 id,  0.0 wa,  0.0 hi,  0.0 si,  0.0 st
KiB Mem :  1863252 total,  1088420 free,   291796 used,   483036 buff/cache
KiB Swap:  2097148 total,  2097148 free,        0 used.  1412036 avail Mem

  PID USER      PR  NI    VIRT    RES    SHR S  %CPU %MEM     TIME+ COMMAND
    7 root      20   0    7312    100      0 R   8.7  0.0   0:59.14 stress
    8 root      20   0    7312    100      0 R   8.7  0.0   0:59.23 stress
   12 root      20   0    7312    100      0 R   8.7  0.0   0:59.23 stress
    6 root      20   0    7312    100      0 R   8.3  0.0   0:59.00 stress
    9 root      20   0    7312    100      0 R   8.3  0.0   0:59.05 stress
   10 root      20   0    7312    100      0 R   8.3  0.0   0:59.09 stress
   11 root      20   0    7312    100      0 R   8.3  0.0   0:59.26 stress
   13 root      20   0    7312    100      0 R   8.3  0.0   0:59.13 stress
   58 root      20   0   56184   1992   1440 R   0.3  0.1   0:00.01 top
    1 root      20   0    7312    632    532 S   0.0  0.0   0:00.04 stress
   44 root      20   0   11804   1884   1488 S   0.0  0.1   0:00.03 bash
```

图 7-6　重新查看 container1 资源占用情况

第五步，设置容器的 CPU 周期和容器在一个周期内可以占用的 CPU 时间，操作命令如下：

[root@docker ~]# docker run -tid --name container3 --cpu-period 100000 --cpu-quota 200000 centos:stress

“--cpu-period”指定容器的 CPU 使用周期，单位是微秒 (μs)，最小值是 1000 μs，最大值是 1 s，默认值是 100 000 μs(0.1 s)。

“--cpu-quota”指定容器在 CPU 使用周期内最多可以占用的时间，单位是微秒 (μs)，默认值是 -1，表示不控制。

以上两个参数一般同时使用，表示容器进程在每个周期中使用单个 CPU 的时间。

[root@docker ~]# docker ps -a

[root@docker ~]# docker exec -it container3 bash

[root@2e04f52d8dfd /]# cat /sys/fs/cgroup/cpu/cpu.cfs_period_us

[root@2e04f52d8dfd /]# cat /sys/fs/cgroup/cpu/cpu.cfs_quota_us

命令运行结果如图 7-7 所示。

```
[root@docker ~]# docker run -tid --name container3 --cpu-period 100000 --cpu-quota 200000 centos:stress
2e04f52d8dfdcad431ed8eefc3cfe9d1f2fa6e66f03be5b96c15f2e745d545eb
[root@docker ~]# docker ps -a
CONTAINER ID   IMAGE           COMMAND       CREATED          STATUS         PORTS    NAMES
2e04f52d8dfd   centos:stress   "/bin/bash"   2 minutes ago    Up 2 minutes            container3
[root@docker ~]# docker exec -it container3 bash
[root@2e04f52d8dfd /]# cat /sys/fs/cgroup/cpu/cpu.cfs_period_us
100000
[root@2e04f52d8dfd /]# cat /sys/fs/cgroup/cpu/cpu.cfs_quota_us
200000
```

图 7-7　设置容器的 CPU 周期和容器在一个周期内可占用的 CPU 时间

7.2.2　限制 CPU 内核、内存和 Block IO

限制 CPU 内核、内存和 Block IO

1. 任务目标

掌握限制 CPU 内核、内存和 Block IO 的方法。

2. 任务内容

(1) 限制 CPU 内核。

(2) 限制 CPU 内存。

(3) 限制 Block IO。

3. 完成任务所需的设备和软件

(1) 一台安装 Windows 10 操作系统的计算机。

(2) VMware Workstation、Docker。

(3) 远程管理工具 MobaXterm。

4. 任务实施步骤

第一步，对于多核服务器，设置容器运行所使用的 CPU 内核，操作命令如下：

```
[root@docker ~]# docker run -tid --name container4 --cpuset-cpus 0-1 centos:stress
```

“--cpuset-cpus”指定容器可以使用的 CPU 内核。

```
[root@docker ~]# docker ps -a
[root@docker ~]# docker exec -it container4 bash
[root@bcb54cf525ef /]# cat /sys/fs/cgroup/cpuset/cpuset.cpus
```

命令运行结果如图 7-8 所示。

```
[root@docker ~]# docker run -tid --name container4 --cpuset-cpus 0-1 centos:stress
bcb54cf525efae098cecb6de7148519c98dc36071778ac95ce15be209276e550
[root@docker ~]# docker ps -a
CONTAINER ID   IMAGE           COMMAND       CREATED         STATUS         PORTS    NAMES
bcb54cf525ef   centos:stress   "/bin/bash"   8 seconds ago   Up 6 seconds            container4
[root@docker ~]# docker exec -it container4 bash
[root@bcb54cf525ef /]# cat /sys/fs/cgroup/cpuset/cpuset.cpus
0-1
```

图 7-8　对于多核服务器设置容器运行所使用的 CPU 内核

第二步，查看容器内所有进程，将容器内进程与 CPU 内核进行绑定，操作命令如下：

```
[root@docker ~]# docker exec  container4 ps aux
[root@docker ~]# docker exec container4 taskset -cp 1
```

其中：

-c：以列表的格式显示和指定 CPU。

-p：在已经存在的 pid 上操作。

1：表示容器内第一个进程编号。

```
[root@docker ~]# docker exec container4 taskset -cp 0 1
```

0：表示第一个 CPU 内核编号。

1：表示第二个 CPU 内核编号。

命令运行结果如图 7-9 所示。

```
[root@docker ~]# docker exec  container4 ps aux
USER        PID %CPU %MEM    VSZ   RSS TTY      STAT START   TIME COMMAND
root          1  0.2  0.0  11828  1652 pts/0    Ss+  10:15   0:00 /bin/bash
root         15  0.0  0.0  51732  1700 ?        Rs   10:15   0:00 ps aux
[root@docker ~]# docker exec container4 taskset -cp 1
pid 1's current affinity list: 0,1
[root@docker ~]# docker exec container4 taskset -cp 0 1
pid 1's current affinity list: 0,1
pid 1's new affinity list: 0
```

图 7-9　将容器进程与 CPU 内核进行绑定

第三步，在同一个 CPU 内核上运行两个容器，分别开启一个 stress 进程，设置两个进程的 CPU 份额为 1∶2，操作命令如下：

```
[root@docker ~]# docker run -tid --name container5 --cpuset-cpus 1 --cpu-shares 521 centos:stress stress -c 1
[root@docker ~]# docker run -tid --name container6 --cpuset-cpus 1 --cpu-shares 1024 centos:stress stress -c 1
[root@docker ~]# docker ps -a
[root@docker ~]# docker exec -it container5 bash
[root@adce81285bc2 /]# top
[root@adce81285bc2 /]# exit
[root@docker ~]# docker exec -it container6 bash
[root@2538d8621bd6 /]# top
```

命令运行结果如图 7-10 和图 7-11 所示。

```
[root@docker ~]# docker run -tid --name container5 --cpuset-cpus 1 --cpu-shares 521 centos:stress stress -c 1
adce81285bc2f64b856bdaf064bfa009a11d5082290e7d4967b69142ac1eb0f2
[root@docker ~]# docker run -tid --name container6 --cpuset-cpus 1 --cpu-shares 1024 centos:stress stress -c 1
2538d8621bd6ca12d1ece6041f08f0bf2a5d9150e8d45cd3736fed498e02ce62
[root@docker ~]# docker ps -a
CONTAINER ID   IMAGE           COMMAND          CREATED              STATUS              PORTS     NAMES
2538d8621bd6   centos:stress   "stress -c 1"    12 seconds ago       Up 11 seconds                 container6
adce81285bc2   centos:stress   "stress -c 1"    About a minute ago   Up About a minute             container5
eb173778423c   centos:stress   "/bin/bash"      About an hour ago    Up About an hour              container4
[root@docker ~]# docker exec -it container5 bash
[root@adce81285bc2 /]# top
top - 11:25:46 up  2:26,  0 users,  load average: 1.62, 0.60, 0.25
Tasks:   4 total,   2 running,   2 sleeping,   0 stopped,   0 zombie
%Cpu(s): 49.7 us,  0.0 sy,  0.0 ni, 49.0 id,  0.0 wa,  0.0 hi,  1.3 si,  0.0 st
KiB Mem :  1863252 total,  1183548 free,   300936 used,   378768 buff/cache
KiB Swap:  2097148 total,  2097148 free,        0 used.  1405184 avail Mem

  PID USER      PR  NI    VIRT    RES    SHR S  %CPU %MEM     TIME+ COMMAND
    7 root      20   0    7312    100      0 R  33.9  0.0   1:37.53 stress
    1 root      20   0    7312    428    344 S   0.0  0.0   0:00.13 stress
    8 root      20   0   11804   1888   1488 S   0.0  0.1   0:00.03 bash
   22 root      20   0   56184   1980   1440 R   0.0  0.1   0:00.43 top
```

图 7-10　在同一个 CPU 内核上运行两个容器

```
[root@adce81285bc2 /]# exit
exit
[root@docker ~]# docker exec -it container6 bash
[root@2538d8621bd6 /]# top
top - 13:10:04 up  4:10,  0 users,  load average: 2.00, 2.01, 2.05
Tasks:   4 total,   2 running,   2 sleeping,   0 stopped,   0 zombie
%Cpu(s): 49.4 us,  0.2 sy,  0.0 ni, 49.1 id,  0.0 wa,  0.0 hi,  1.3 si,  0.0 st
KiB Mem :  1863252 total,  1182824 free,   301268 used,   379160 buff/cache
KiB Swap:  2097148 total,  2097148 free,        0 used.  1404776 avail Mem

  PID USER      PR  NI    VIRT    RES    SHR S  %CPU %MEM     TIME+ COMMAND
    7 root      20   0    7312    100      0 R  66.1  0.0  69:49.33 stress
    1 root      20   0    7312    428    344 S   0.0  0.0   0:00.04 stress
   25 root      20   0   11828   1888   1488 S   0.0  0.1   0:00.03 bash
   39 root      20   0   56184   1980   1440 R   0.0  0.1   0:00.00 top
```

图 7-11　在同一个 CPU 内核上运行两个容器

从图中可以看出，第二个 CPU 内核的资源已使用完，使用率为 100%。两容器的 CPU

份额为1∶2。

第四步，拉取对容器执行压力测试的镜像progrium/stress，分配容器内存，操作命令如下：

```
[root@docker ~]# docker run -it --name container7 -m 128M --memory-swap=256M progrium/stress --vm 1 --vm-bytes 220M
```

其中：

-m或--memory：用于设置内存的使用限额。

--memory-swap：用于设置交换分区的使用限额。

--vm 1：用于启动一个内存工作线程。

--vm-bytes 220M：用于为每个线程分配220 MB内存。

命令运行结果如图7-12所示。

```
[root@docker ~]# docker run -it --name container7 -m 128M --memory-swap=256M progrium/stress --vm 1 --vm-bytes 220M
stress: info: [1] dispatching hogs: 0 cpu, 0 io, 1 vm, 0 hdd
stress: dbug: [1] using backoff sleep of 3000us
stress: dbug: [1] --> hogvm worker 1 [7] forked
stress: dbug: [7] allocating 230686720 bytes ...
stress: dbug: [7] touching bytes in strides of 4096 bytes ...
stress: dbug: [7] freed 230686720 bytes
stress: dbug: [7] allocating 230686720 bytes ...
stress: dbug: [7] touching bytes in strides of 4096 bytes ...
stress: dbug: [7] freed 230686720 bytes
stress: dbug: [7] allocating 230686720 bytes ...
stress: dbug: [7] touching bytes in strides of 4096 bytes ...
stress: dbug: [7] freed 230686720 bytes
stress: dbug: [7] allocating 230686720 bytes ...
stress: dbug: [7] touching bytes in strides of 4096 bytes ...
stress: dbug: [7] freed 230686720 bytes
stress: dbug: [7] allocating 230686720 bytes ...
stress: dbug: [7] touching bytes in strides of 4096 bytes ...
stress: dbug: [7] freed 230686720 bytes
stress: dbug: [7] allocating 230686720 bytes ...
```

图7-12　分配容器内存，线程工作正常

分配给线程的220 MB内存小于256 MB，从图中可见，工作线程运行正常。当分配给线程的内存为300 MB时，操作命令如下：

```
[root@docker ~]# docker run -it --name container8 -m 128M --memory-swap=256M progrium/stress --vm 1 --vm-bytes 300M
```

命令运行结果如图7-13所示。

```
[root@docker ~]# docker run -it --name container8 -m 128M --memory-swap=256M progrium/stress --vm 1 --vm-bytes 300M
stress: info: [1] dispatching hogs: 0 cpu, 0 io, 1 vm, 0 hdd
stress: dbug: [1] using backoff sleep of 3000us
stress: dbug: [1] --> hogvm worker 1 [8] forked
stress: dbug: [8] allocating 314572800 bytes ...
stress: dbug: [8] touching bytes in strides of 4096 bytes ...
stress: FAIL: [1] (416) <-- worker 8 got signal 9
stress: WARN: [1] (418) now reaping child worker processes
stress: FAIL: [1] (422) kill error: No such process
stress: FAIL: [1] (452) failed run completed in 1s
```

图7-13　分配容器内存，线程报错

从图中可见，当分配给线程的内存为300 MB大于256 MB时，线程报错，容器退出。

第五步，先查看容器写速率，然后设置容器写速率为10 MB/s，再查看是否设置成功，操作命令如下：

```
[root@docker ~]# docker run -it --name container11 centos:stress
[root@36b8a569842a /]# dd if=/dev/zero of=/tmp/test bs=2M count=30 oflag=direct
```

其中：

if：输入文件 (从 /dev/zero 读取无限的零，该文件是一个字符设备文件)。

of：输出文件 (写入 /tmp/test)。

bs：表示块大小，设置为 2 M。

count：表示要复制的块数，设置为 128。

oflag=direct：表示以直接 IO 模式进行操作。

[root@36b8a569842a /]# exit

[root@docker ~]# docker run -it --name container12 --device-write-bps /dev/sda:10MB centos:stress

--device-write-bps：表示限制写某个设备的速率，单位为 bps(每秒读写的数据量)。

[root@1fd86d5092e6 /]# dd if=/dev/zero of=/tmp/test bs=2M count=30 oflag=direct

命令运行结果如图 7-14 所示。

```
[root@docker ~]# docker run -it --name container11 centos:stress
[root@36b8a569842a /]# dd if=/dev/zero of=/tmp/test bs=2M count=30 oflag=direct
30+0 records in
30+0 records out
62914560 bytes (63 MB) copied, 0.0963755 s, 653 MB/s
[root@36b8a569842a /]# exit
exit
[root@docker ~]# docker run -it --name container12 --device-write-bps /dev/sda:10MB centos:stress
[root@1fd86d5092e6 /]# dd if=/dev/zero of=/tmp/test bs=2M count=30 oflag=direct
30+0 records in
30+0 records out
62914560 bytes (63 MB) copied, 5.95178 s, 10.6 MB/s
```

图 7-14　设置容器写速率

7.2.3　查看 Docker 日志

查看 Docker 日志

1. 任务目标

掌握查看 Docker 日志管理的方法。

2. 任务内容

(1) 查看 Docker 引擎日志。

(2) 查看 Docker 容器的最新日志。

(3) 实时查看与查看指定行数的 Docker 容器日志。

(4) 查看指定时间的 Docker 容器日志。

(5) 查看根据关键词过滤的 Docker 容器日志。

(6) 将 Docker 容器日志输出到文件。

3. 完成任务所需的设备和软件

(1) 一台安装 Windows 10 操作系统的计算机。

(2) VMware Workstation、Docker。

(3) 远程管理工具 MobaXterm。

4. 任务实施步骤

第一步，查看 Docker 引擎日志，操作命令如下：

```
[root@docker ~]# journalctl -u docker
```

journalctl：systemd 的日志查询工具，用于查询 systemd 及其服务的日志。

“-u”用于指定服务。

命令运行结果如图 7-15 所示。

```
[root@docker ~]# journalctl -u docker
-- Logs begin at Mon 2024-08-12 07:01:09 CST, end at Mon 2024-08-12 10:05:16 CST. --
Aug 12 07:02:25 master systemd[1]: Starting Docker Application Container Engine...
Aug 12 07:03:03 master dockerd[8621]: time="2024-08-12T07:03:03.650918601+08:00" level=info msg="Starting up"
Aug 12 07:03:03 master dockerd[8621]: time="2024-08-12T07:03:03.710077883+08:00" level=info msg="parsed scheme: \"unix\"" module=grpc
Aug 12 07:03:03 master dockerd[8621]: time="2024-08-12T07:03:03.718429685+08:00" level=info msg="scheme \"unix\" not registered, fallback
Aug 12 07:03:03 master dockerd[8621]: time="2024-08-12T07:03:03.721102257+08:00" level=info msg="ccResolverWrapper: sending update to cc:
Aug 12 07:03:03 master dockerd[8621]: time="2024-08-12T07:03:03.721223661+08:00" level=info msg="ClientConn switching balancer to \"pick_f
Aug 12 07:03:03 master dockerd[8621]: time="2024-08-12T07:03:03.851041634+08:00" level=info msg="parsed scheme: \"unix\"" module=grpc
Aug 12 07:03:03 master dockerd[8621]: time="2024-08-12T07:03:03.851092150+08:00" level=info msg="scheme \"unix\" not registered, fallback
Aug 12 07:03:03 master dockerd[8621]: time="2024-08-12T07:03:03.851130336+08:00" level=info msg="ccResolverWrapper: sending update to cc:
Aug 12 07:03:03 master dockerd[8621]: time="2024-08-12T07:03:03.851152085+08:00" level=info msg="ClientConn switching balancer to \"pick_f
Aug 12 07:03:08 master dockerd[8621]: time="2024-08-12T07:03:08.083196290+08:00" level=info msg="[graphdriver] using prior storage driver:
Aug 12 07:03:21 master dockerd[8621]: time="2024-08-12T07:03:21.492137391+08:00" level=info msg="Loading containers: start."
Aug 12 07:03:24 master dockerd[8621]: time="2024-08-12T07:03:24.421222660+08:00" level=info msg="Default bridge (docker0) is assigned with
Aug 12 07:03:24 master dockerd[8621]: time="2024-08-12T07:03:24.614129347+08:00" level=info msg="Loading containers: done."
Aug 12 07:03:25 master dockerd[8621]: time="2024-08-12T07:03:25.138708384+08:00" level=info msg="Docker daemon" commit=87a90dc graphdriver
Aug 12 07:03:25 master dockerd[8621]: time="2024-08-12T07:03:25.144268427+08:00" level=info msg="Daemon has completed initialization"
Aug 12 07:03:25 master systemd[1]: Started Docker Application Container Engine.
Aug 12 07:03:25 master dockerd[8621]: time="2024-08-12T07:03:25.328535393+08:00" level=info msg="API listen on /var/run/docker.sock"
Aug 12 07:04:05 master dockerd[8621]: time="2024-08-12T07:04:05.260551444+08:00" level=info msg="ignoring event" container=35656fbcec0eadf
Aug 12 07:04:05 master dockerd[8621]: time="2024-08-12T07:04:05.504925384+08:00" level=info msg="ignoring event" container=db1d7ec6f057759
Aug 12 07:04:08 master dockerd[8621]: time="2024-08-12T07:04:08.469702116+08:00" level=info msg="ignoring event" container=881bdef6a63d2c5
Aug 12 07:04:08 master dockerd[8621]: time="2024-08-12T07:04:08.498025700+08:00" level=info msg="ignoring event" container=23e2ed427438c2c
Aug 12 07:04:11 master dockerd[8621]: time="2024-08-12T07:04:11.114014292+08:00" level=info msg="ignoring event" container=05f65e419b0eac4
Aug 12 07:04:11 master dockerd[8621]: time="2024-08-12T07:04:11.266579592+08:00" level=info msg="ignoring event" container=3880a6d17824fe6
Aug 12 07:04:13 master dockerd[8621]: time="2024-08-12T07:04:13.365419742+08:00" level=info msg="ignoring event" container=6e898994d591066
Aug 12 07:04:13 master dockerd[8621]: time="2024-08-12T07:04:13.447340767+08:00" level=info msg="ignoring event" container=41c307cf6dda653
Aug 12 07:04:15 master dockerd[8621]: time="2024-08-12T07:04:15.523567051+08:00" level=info msg="ignoring event" container=5d71258a96f0d7b
Aug 12 07:04:15 master dockerd[8621]: time="2024-08-12T07:04:15.555372408+08:00" level=info msg="ignoring event" container=314cdc06905b9c5
Aug 12 07:04:17 master dockerd[8621]: time="2024-08-12T07:04:17.522610326+08:00" level=info msg="ignoring event" container=9bf7a4a119fec33
Aug 12 07:04:17 master dockerd[8621]: time="2024-08-12T07:04:17.576329926+08:00" level=info msg="ignoring event" container=9f02feb5d6a95e7
Aug 12 07:04:18 master dockerd[8621]: time="2024-08-12T07:04:18.721676113+08:00" level=info msg="ignoring event" container=67b2347a681f8b2
Aug 12 07:04:18 master dockerd[8621]: time="2024-08-12T07:04:18.854258047+08:00" level=info msg="ignoring event" container=66742be22448f78
Aug 12 07:04:19 master dockerd[8621]: time="2024-08-12T07:04:19.762086740+08:00" level=info msg="ignoring event" container=f1e3c335c94dbb0
Aug 12 07:04:20 master dockerd[8621]: time="2024-08-12T07:04:20.868624569+08:00" level=info msg="ignoring event" container=35dbe1438f45170
Aug 12 07:04:21 master dockerd[8621]: time="2024-08-12T07:04:21.068736857+08:00" level=info msg="ignoring event" container=02a3a0341a55700
Aug 12 07:04:21 master dockerd[8621]: time="2024-08-12T07:04:21.983242639+08:00" level=info msg="ignoring event" container=06e84a48ac93051
Aug 12 07:04:23 master dockerd[8621]: time="2024-08-12T07:04:23.265536165+08:00" level=info msg="ignoring event" container=e8810917c3fe880
```

图 7-15 查看 Docker 引擎日志

第二步，查看 Docker 容器的日志，操作命令如下：

```
[root@docker ~]# docker logs ea878a2c18c9( 容器 ID 或容器名称 )
```

命令运行结果如图 7-16 所示。

```
[root@docker ~]# docker logs ea878a2c18c9
/docker-entrypoint.sh: /docker-entrypoint.d/ is not empty, will attempt to perform configuration
/docker-entrypoint.sh: Looking for shell scripts in /docker-entrypoint.d/
/docker-entrypoint.sh: Launching /docker-entrypoint.d/10-listen-on-ipv6-by-default.sh
10-listen-on-ipv6-by-default.sh: info: Getting the checksum of /etc/nginx/conf.d/default.conf
10-listen-on-ipv6-by-default.sh: info: Enabled listen on IPv6 in /etc/nginx/conf.d/default.conf
/docker-entrypoint.sh: Sourcing /docker-entrypoint.d/15-local-resolvers.envsh
/docker-entrypoint.sh: Launching /docker-entrypoint.d/20-envsubst-on-templates.sh
/docker-entrypoint.sh: Launching /docker-entrypoint.d/30-tune-worker-processes.sh
/docker-entrypoint.sh: Configuration complete; ready for start up
2024/08/09 03:22:30 [notice] 1#1: using the "epoll" event method
2024/08/09 03:22:30 [notice] 1#1: nginx/1.25.3
2024/08/09 03:22:30 [notice] 1#1: built by gcc 12.2.0 (Debian 12.2.0-14)
2024/08/09 03:22:30 [notice] 1#1: OS: Linux 3.10.0-957.el7.x86_64
2024/08/09 03:22:30 [notice] 1#1: getrlimit(RLIMIT_NOFILE): 65536:65536
2024/08/09 03:22:30 [notice] 1#1: start worker processes
2024/08/09 03:22:30 [notice] 1#1: start worker process 29
2024/08/09 03:22:30 [notice] 1#1: start worker process 30
192.168.1.1 - - [09/Aug/2024:03:23:03 +0000] "GET / HTTP/1.1" 200 615 "-" "Mozilla/5.0 (Windows NT 6.1; Win64; x64; rv:109.0) Gecko/201001
01 Firefox/115.0" "-"
2024/08/09 03:23:03 [error] 29#29: *1 open() "/usr/share/nginx/html/favicon.ico" failed (2: No such file or directory), client: 192.168.1.
1, server: localhost, request: "GET /favicon.ico HTTP/1.1", host: "192.168.1.10:8002", referrer: "http://192.168.1.10:8002/"
192.168.1.1 - - [09/Aug/2024:03:23:03 +0000] "GET /favicon.ico HTTP/1.1" 404 153 "http://192.168.1.10:8002/" "Mozilla/5.0 (Windows NT 6.1;
 Win64; x64; rv:109.0) Gecko/20100101 Firefox/115.0" "-"
2024/08/09 04:51:31 [notice] 1#1: signal 3 (SIGQUIT) received, shutting down
2024/08/09 04:51:31 [notice] 30#30: gracefully shutting down
2024/08/09 04:51:31 [notice] 30#30: exiting
2024/08/09 04:51:31 [notice] 29#29: gracefully shutting down
2024/08/09 04:51:31 [notice] 29#29: exiting
2024/08/09 04:51:31 [notice] 29#29: exit
2024/08/09 04:51:31 [notice] 30#30: exit
2024/08/09 04:51:31 [notice] 1#1: signal 17 (SIGCHLD) received from 29
2024/08/09 04:51:31 [notice] 1#1: worker process 29 exited with code 0
2024/08/09 04:51:31 [notice] 1#1: signal 29 (SIGIO) received
2024/08/09 04:51:31 [notice] 1#1: signal 17 (SIGCHLD) received from 30
2024/08/09 04:51:31 [notice] 1#1: worker process 30 exited with code 0
2024/08/09 04:51:31 [notice] 1#1: exit
[root@docker ~]#
```

图 7-16 查看 Docker 容器的日志

第三步，实时查看 Docker 容器的日志，操作命令如下：

[root@docker ~]# docker logs –f b2c37cee717c(容器 ID 或容器名称)

“-f”用于实时跟踪容器的日志输出。

命令运行结果如图 7-17 所示。

```
[root@docker ~]# docker logs -f b2c37cee717c
Flag --insecure-port has been deprecated, This flag will be removed in a future version.
I0811 14:16:17.141286       1 server.go:625] external host was not specified, using 192.168.1.20
I0811 14:16:17.142002       1 server.go:163] Version: v1.19.0
I0811 14:16:18.458420       1 plugins.go:158] Loaded 12 mutating admission controller(s) successfully in the following order: NamespaceLif
ecycle,LimitRanger,ServiceAccount,NodeRestriction,TaintNodesByCondition,Priority,DefaultTolerationSeconds,DefaultStorageClass,StorageObjec
tInUseProtection,RuntimeClass,DefaultIngressClass,MutatingAdmissionWebhook.
I0811 14:16:18.458446       1 plugins.go:161] Loaded 10 validating admission controller(s) successfully in the following order: LimitRange
r,ServiceAccount,Priority,PersistentVolumeClaimResize,RuntimeClass,CertificateApproval,CertificateSigning,CertificateSubjectRestriction,Va
lidatingAdmissionWebhook,ResourceQuota.
I0811 14:16:18.459883       1 plugins.go:158] Loaded 12 mutating admission controller(s) successfully in the following order: NamespaceLif
ecycle,LimitRanger,ServiceAccount,NodeRestriction,TaintNodesByCondition,Priority,DefaultTolerationSeconds,DefaultStorageClass,StorageObjec
tInUseProtection,RuntimeClass,DefaultIngressClass,MutatingAdmissionWebhook.
I0811 14:16:18.459905       1 plugins.go:161] Loaded 10 validating admission controller(s) successfully in the following order: LimitRange
r,ServiceAccount,Priority,PersistentVolumeClaimResize,RuntimeClass,CertificateApproval,CertificateSigning,CertificateSubjectRestriction,Va
lidatingAdmissionWebhook,ResourceQuota.
I0811 14:16:18.549978       1 client.go:360] parsed scheme: "endpoint"
I0811 14:16:18.550085       1 endpoint.go:68] ccResolverWrapper: sending new addresses to cc: [{https://127.0.0.1:2379  <nil> 0 <nil>}]
W0811 14:16:18.557359       1 clientconn.go:1223] grpc: addrConn.createTransport failed to connect to {https://127.0.0.1:2379  <nil> 0 <ni
l>}. Err :connection error: desc = "transport: Error while dialing dial tcp 127.0.0.1:2379: connect: connection refused". Reconnecting...
I0811 14:16:19.409127       1 client.go:360] parsed scheme: "endpoint"
I0811 14:16:19.409167       1 endpoint.go:68] ccResolverWrapper: sending new addresses to cc: [{https://127.0.0.1:2379  <nil> 0 <nil>}]
I0811 14:16:24.191986       1 client.go:360] parsed scheme: "endpoint"
I0811 14:16:24.192033       1 endpoint.go:68] ccResolverWrapper: sending new addresses to cc: [{https://127.0.0.1:2379  <nil> 0 <nil>}]
I0811 14:16:24.219654       1 client.go:360] parsed scheme: "endpoint"
I0811 14:16:24.219712       1 endpoint.go:68] ccResolverWrapper: sending new addresses to cc: [{https://127.0.0.1:2379  <nil> 0 <nil>}]
I0811 14:16:24.221249       1 client.go:360] parsed scheme: "passthrough"
I0811 14:16:24.221424       1 passthrough.go:48] ccResolverWrapper: sending update to cc: {[{https://127.0.0.1:2379  <nil> 0 <nil>}] <nil>
 <nil>}
I0811 14:16:24.221442       1 clientconn.go:948] ClientConn switching balancer to "pick_first"
I0811 14:16:24.391838       1 master.go:271] Using reconciler: lease
I0811 14:16:24.392443       1 client.go:360] parsed scheme: "endpoint"
I0811 14:16:24.392470       1 endpoint.go:68] ccResolverWrapper: sending new addresses to cc: [{https://127.0.0.1:2379  <nil> 0 <nil>}]
I0811 14:16:24.421236       1 client.go:360] parsed scheme: "endpoint"
I0811 14:16:24.421311       1 endpoint.go:68] ccResolverWrapper: sending new addresses to cc: [{https://127.0.0.1:2379  <nil> 0 <nil>}]
I0811 14:16:24.461076       1 client.go:360] parsed scheme: "endpoint"
```

图 7-17　实时查看 Docker 容器的最新日志

第四步，查看指定行数的 Docker 容器日志，操作命令如下：

[root@docker ~]# docker logs --tail 20 b2c37cee717c

“--tail”用于指定查看的日志行数。

命令运行结果如图 7-18 所示。

```
[root@docker ~]# docker logs --tail 20 b2c37cee717c
I0811 14:23:38.041672       1 passthrough.go:48] ccResolverWrapper: sending update to cc: {[{https://127.0.0.1:2379  <nil> 0 <nil>}] <nil>
 <nil>}
I0811 14:23:38.041685       1 clientconn.go:948] ClientConn switching balancer to "pick_first"
I0811 14:24:12.128985       1 client.go:360] parsed scheme: "passthrough"
I0811 14:24:12.129054       1 passthrough.go:48] ccResolverWrapper: sending update to cc: {[{https://127.0.0.1:2379  <nil> 0 <nil>}] <nil>
 <nil>}
I0811 14:24:12.129086       1 clientconn.go:948] ClientConn switching balancer to "pick_first"
I0811 14:24:46.182711       1 client.go:360] parsed scheme: "passthrough"
I0811 14:24:46.182841       1 passthrough.go:48] ccResolverWrapper: sending update to cc: {[{https://127.0.0.1:2379  <nil> 0 <nil>}] <nil>
 <nil>}
I0811 14:24:46.182865       1 clientconn.go:948] ClientConn switching balancer to "pick_first"
I0811 14:25:23.847784       1 client.go:360] parsed scheme: "passthrough"
I0811 14:25:23.847889       1 passthrough.go:48] ccResolverWrapper: sending update to cc: {[{https://127.0.0.1:2379  <nil> 0 <nil>}] <nil>
 <nil>}
I0811 14:25:23.847902       1 clientconn.go:948] ClientConn switching balancer to "pick_first"
I0811 14:25:56.574694       1 client.go:360] parsed scheme: "passthrough"
I0811 14:25:56.574751       1 passthrough.go:48] ccResolverWrapper: sending update to cc: {[{https://127.0.0.1:2379  <nil> 0 <nil>}] <nil>
 <nil>}
I0811 14:25:56.574762       1 clientconn.go:948] ClientConn switching balancer to "pick_first"
I0811 14:26:34.751123       1 client.go:360] parsed scheme: "passthrough"
I0811 14:26:34.751187       1 passthrough.go:48] ccResolverWrapper: sending update to cc: {[{https://127.0.0.1:2379  <nil> 0 <nil>}] <nil>
 <nil>}
I0811 14:26:34.751199       1 clientconn.go:948] ClientConn switching balancer to "pick_first"
I0811 14:27:16.869736       1 client.go:360] parsed scheme: "passthrough"
I0811 14:27:16.869801       1 passthrough.go:48] ccResolverWrapper: sending update to cc: {[{https://127.0.0.1:2379  <nil> 0 <nil>}] <nil>
 <nil>}
I0811 14:27:16.869852       1 clientconn.go:948] ClientConn switching balancer to "pick_first"
[root@master ~]# █
```

图 7-18　查看指定行数的 Docker 容器日志

第五步，查看 Docker 容器日志的时间戳，操作命令如下：

[root@docker ~]# docker logs --timestamps b2c37cee717c

“--timestamps”用于在日志中显示时间戳。

命令运行结果如图 7-19 所示。

```
[root@docker ~]# docker logs --timestamps b2c37cee717c
2024-08-11T14:16:17.156574613Z Flag --insecure-port has been deprecated, This flag will be removed in a future version.
2024-08-11T14:16:17.156666347Z I0811 14:16:17.141286       1 server.go:625] external host was not specified, using 192.168.1.20
2024-08-11T14:16:17.156676231Z I0811 14:16:17.142002       1 server.go:163] Version: v1.19.0
2024-08-11T14:16:18.462226038Z I0811 14:16:18.458420       1 plugins.go:158] Loaded 12 mutating admission controller(s) successfully in th
e following order: NamespaceLifecycle,LimitRanger,ServiceAccount,NodeRestriction,TaintNodesByCondition,Priority,DefaultTolerationSeconds,D
efaultStorageClass,StorageObjectInUseProtection,RuntimeClass,DefaultIngressClass,MutatingAdmissionWebhook.
2024-08-11T14:16:18.462260780Z I0811 14:16:18.458446       1 plugins.go:161] Loaded 10 validating admission controller(s) successfully in
the following order: LimitRanger,ServiceAccount,Priority,PersistentVolumeClaimResize,RuntimeClass,CertificateApproval,CertificateSigning,C
ertificateSubjectRestriction,ValidatingAdmissionWebhook,ResourceQuota.
2024-08-11T14:16:18.462268040Z I0811 14:16:18.459883       1 plugins.go:158] Loaded 12 mutating admission controller(s) successfully in th
e following order: NamespaceLifecycle,LimitRanger,ServiceAccount,NodeRestriction,TaintNodesByCondition,Priority,DefaultTolerationSeconds,D
efaultStorageClass,StorageObjectInUseProtection,RuntimeClass,DefaultIngressClass,MutatingAdmissionWebhook.
2024-08-11T14:16:18.462272537Z I0811 14:16:18.459905       1 plugins.go:161] Loaded 10 validating admission controller(s) successfully in
the following order: LimitRanger,ServiceAccount,Priority,PersistentVolumeClaimResize,RuntimeClass,CertificateApproval,CertificateSigning,C
ertificateSubjectRestriction,ValidatingAdmissionWebhook,ResourceQuota.
2024-08-11T14:16:18.551504668Z I0811 14:16:18.549978       1 client.go:360] parsed scheme: "endpoint"
2024-08-11T14:16:18.551537169Z I0811 14:16:18.550085       1 endpoint.go:68] ccResolverWrapper: sending new addresses to cc: [{https://127
.0.0.1:2379  <nil> 0 <nil>}]
2024-08-11T14:16:18.559467630Z W0811 14:16:18.557359       1 clientconn.go:1223] grpc: addrConn.createTransport failed to connect to {http
s://127.0.0.1:2379  <nil> 0 <nil>}. Err :connection error: desc = "transport: Error while dialing dial tcp 127.0.0.1:2379: connect: connec
tion refused". Reconnecting...
2024-08-11T14:16:19.410005636Z I0811 14:16:19.409127       1 client.go:360] parsed scheme: "endpoint"
2024-08-11T14:16:19.410039619Z I0811 14:16:19.409167       1 endpoint.go:68] ccResolverWrapper: sending new addresses to cc: [{https://127
.0.0.1:2379  <nil> 0 <nil>}]
2024-08-11T14:16:24.192212324Z I0811 14:16:24.191986       1 client.go:360] parsed scheme: "endpoint"
2024-08-11T14:16:24.192245117Z I0811 14:16:24.192033       1 endpoint.go:68] ccResolverWrapper: sending new addresses to cc: [{https://127
.0.0.1:2379  <nil> 0 <nil>}]
2024-08-11T14:16:24.220806257Z I0811 14:16:24.219654       1 client.go:360] parsed scheme: "endpoint"
2024-08-11T14:16:24.220840958Z I0811 14:16:24.219712       1 endpoint.go:68] ccResolverWrapper: sending new addresses to cc: [{https://127
.0.0.1:2379  <nil> 0 <nil>}]
2024-08-11T14:16:24.222294486Z I0811 14:16:24.221249       1 client.go:360] parsed scheme: "passthrough"
2024-08-11T14:16:24.222324547Z I0811 14:16:24.221424       1 passthrough.go:48] ccResolverWrapper: sending update to cc: {[{https://127.0.
0.1:2379  <nil> 0 <nil>}] <nil> <nil>}
2024-08-11T14:16:24.222342207Z I0811 14:16:24.221442       1 clientconn.go:948] ClientConn switching balancer to "pick_first"
2024-08-11T14:16:24.392957753Z I0811 14:16:24.391838       1 master.go:271] Using reconciler: lease
2024-08-11T14:16:24.392990576Z I0811 14:16:24.392443       1 client.go:360] parsed scheme: "endpoint"
2024-08-11T14:16:24.392997376Z I0811 14:16:24.392470       1 endpoint.go:68] ccResolverWrapper: sending new addresses to cc: [{https://127
.0.0.1:2379  <nil> 0 <nil>}]
```

图 7-19　查看 Docker 容器日志的时间戳

第六步，查看在指定时间之后的 Docker 容器日志，操作命令如下：

```
[root@docker ~]# docker logs --since '2024-05-10T06:00:00' b2c37cee717c
```

“--since”用于查看指定时间之后的日志。

命令运行结果如图 7-20 所示。

```
[root@docker ~]# docker logs --since '2024-05-10T06:00:00' b2c37cee717c
Flag --insecure-port has been deprecated, This flag will be removed in a future version.
I0811 14:16:17.141286       1 server.go:625] external host was not specified, using 192.168.1.20
I0811 14:16:17.142002       1 server.go:163] Version: v1.19.0
I0811 14:16:18.458420       1 plugins.go:158] Loaded 12 mutating admission controller(s) successfully in the following order: NamespaceLif
ecycle,LimitRanger,ServiceAccount,NodeRestriction,TaintNodesByCondition,Priority,DefaultTolerationSeconds,DefaultStorageClass,StorageObjec
tInUseProtection,RuntimeClass,DefaultIngressClass,MutatingAdmissionWebhook.
I0811 14:16:18.458446       1 plugins.go:161] Loaded 10 validating admission controller(s) successfully in the following order: LimitRange
r,ServiceAccount,Priority,PersistentVolumeClaimResize,RuntimeClass,CertificateApproval,CertificateSigning,CertificateSubjectRestriction,Va
lidatingAdmissionWebhook,ResourceQuota.
I0811 14:16:18.459883       1 plugins.go:158] Loaded 12 mutating admission controller(s) successfully in the following order: NamespaceLif
ecycle,LimitRanger,ServiceAccount,NodeRestriction,TaintNodesByCondition,Priority,DefaultTolerationSeconds,DefaultStorageClass,StorageObjec
tInUseProtection,RuntimeClass,DefaultIngressClass,MutatingAdmissionWebhook.
I0811 14:16:18.459905       1 plugins.go:161] Loaded 10 validating admission controller(s) successfully in the following order: LimitRange
r,ServiceAccount,Priority,PersistentVolumeClaimResize,RuntimeClass,CertificateApproval,CertificateSigning,CertificateSubjectRestriction,Va
lidatingAdmissionWebhook,ResourceQuota.
I0811 14:16:18.549978       1 client.go:360] parsed scheme: "endpoint"
I0811 14:16:18.550085       1 endpoint.go:68] ccResolverWrapper: sending new addresses to cc: [{https://127.0.0.1:2379  <nil> 0 <nil>}]
W0811 14:16:18.557359       1 clientconn.go:1223] grpc: addrConn.createTransport failed to connect to {https://127.0.0.1:2379  <nil> 0 <ni
l>}. Err :connection error: desc = "transport: Error while dialing dial tcp 127.0.0.1:2379: connect: connection refused". Reconnecting...
I0811 14:16:19.409127       1 client.go:360] parsed scheme: "endpoint"
I0811 14:16:19.409167       1 endpoint.go:68] ccResolverWrapper: sending new addresses to cc: [{https://127.0.0.1:2379  <nil> 0 <nil>}]
I0811 14:16:24.191986       1 client.go:360] parsed scheme: "endpoint"
I0811 14:16:24.192033       1 endpoint.go:68] ccResolverWrapper: sending new addresses to cc: [{https://127.0.0.1:2379  <nil> 0 <nil>}]
I0811 14:16:24.219654       1 client.go:360] parsed scheme: "endpoint"
I0811 14:16:24.219712       1 endpoint.go:68] ccResolverWrapper: sending new addresses to cc: [{https://127.0.0.1:2379  <nil> 0 <nil>}]
I0811 14:16:24.221249       1 client.go:360] parsed scheme: "passthrough"
I0811 14:16:24.221424       1 passthrough.go:48] ccResolverWrapper: sending update to cc: {[{https://127.0.0.1:2379  <nil> 0 <nil>}] <nil>
 <nil>}
I0811 14:16:24.221442       1 clientconn.go:948] ClientConn switching balancer to "pick_first"
I0811 14:16:24.391838       1 master.go:271] Using reconciler: lease
I0811 14:16:24.392443       1 client.go:360] parsed scheme: "endpoint"
I0811 14:16:24.392470       1 endpoint.go:68] ccResolverWrapper: sending new addresses to cc: [{https://127.0.0.1:2379  <nil> 0 <nil>}]
I0811 14:16:24.421236       1 client.go:360] parsed scheme: "endpoint"
I0811 14:16:24.421311       1 endpoint.go:68] ccResolverWrapper: sending new addresses to cc: [{https://127.0.0.1:2379  <nil> 0 <nil>}]
I0811 14:16:24.461076       1 client.go:360] parsed scheme: "endpoint"
I0811 14:16:24.461106       1 endpoint.go:68] ccResolverWrapper: sending new addresses to cc: [{https://127.0.0.1:2379  <nil> 0 <nil>}]
I0811 14:16:24.475102       1 client.go:360] parsed scheme: "endpoint"
I0811 14:16:24.475130       1 endpoint.go:68] ccResolverWrapper: sending new addresses to cc: [{https://127.0.0.1:2379  <nil> 0 <nil>}]
```

图 7-20　查看在指定时间之后的 Docker 容器日志

第七步，查看最近 30 分钟的 Docker 容器日志，操作命令如下：

```
[root@docker ~]# docker logs --since 30m be0addc546d6
```

命令运行结果如图 7-21 所示。

```
[root@docker ~]# docker logs --since 30m be0addc546d6
Welcome
Welcome
Welcome
Welcome
Welcome
Welcome
Welcome
Welcome
Welcome
Welcome
Welcome
Welcome
Welcome
Welcome
Welcome
Welcome
Welcome
Welcome
Welcome
Welcome
Welcome
Welcome
Welcome
Welcome
Welcome
Welcome
```

图 7-21　查看最近 30 分钟的 Docker 容器日志

第八步，查看在指定时间之前的 Docker 容器日志，操作命令如下：

```
[root@docker ~]# docker logs --until '2024-06-10T06:00:00' ee6dba75e87d
```

“--until”用于查看指定时间之前的日志。

命令运行结果如图 7-22 所示。

```
[root@docker ~]# docker logs --until '2024-06-10T06:00:00' ee6dba75e87d
I0601 23:09:24.230724       1 main.go:212] CLI flags config: {etcdEndpoints:http://127.0.0.1:4001,http://127.0.0.1:2379 etcdPrefix:/coreos
.com/network etcdKeyfile: etcdCertfile: etcdCAFile: etcdUsername: etcdPassword: version:false kubeSubnetMgr:true kubeApiUrl: kubeAnnotatio
nPrefix:flannel.alpha.coreos.com kubeConfigFile: iface:[] ifaceRegex:[] ipMasq:true ifaceCanReach: subnetFile:/run/flannel/subnet.env publ
icIP: publicIPv6: subnetLeaseRenewMargin:60 healthzIP:0.0.0.0 healthzPort:0 iptablesResyncSeconds:5 iptablesForwardRules:true netConfPath:
/etc/kube-flannel/net-conf.json setNodeNetworkUnavailable:true useMultiClusterCidr:false}
W0601 23:09:24.231140       1 client_config.go:617] Neither --kubeconfig nor --master was specified.  Using the inClusterConfig.  This mig
ht not work.
I0601 23:09:24.438878       1 kube.go:145] Waiting 10m0s for node controller to sync
I0601 23:09:24.439118       1 kube.go:490] Starting kube subnet manager
I0601 23:09:24.503183       1 kube.go:511] Creating the node lease for IPv4. This is the n.Spec.PodCIDRs: [10.244.0.0/24]
I0601 23:09:24.503429       1 kube.go:511] Creating the node lease for IPv4. This is the n.Spec.PodCIDRs: [10.244.1.0/24]
I0601 23:09:24.503492       1 kube.go:511] Creating the node lease for IPv4. This is the n.Spec.PodCIDRs: [10.244.2.0/24]
I0601 23:09:25.439274       1 kube.go:152] Node controller sync successful
I0601 23:09:25.439315       1 main.go:232] Created subnet manager: Kubernetes Subnet Manager - node1
I0601 23:09:25.439334       1 main.go:235] Installing signal handlers
I0601 23:09:25.440425       1 main.go:543] Found network config - Backend type: vxlan
I0601 23:09:25.440717       1 match.go:206] Determining IP address of default interface
I0601 23:09:25.443927       1 match.go:259] Using interface with name ens33 and address 192.168.1.21
I0601 23:09:25.446787       1 match.go:281] Defaulting external address to interface address (192.168.1.21)
I0601 23:09:25.447144       1 vxlan.go:141] VXLAN config: VNI=1 Port=0 GBP=false Learning=false DirectRouting=false
W0601 23:09:25.537780       1 main.go:596] no subnet found for key: FLANNEL_SUBNET in file: /run/flannel/subnet.env
I0601 23:09:25.537808       1 main.go:482] Current network or subnet (10.244.0.0/16, 10.244.1.0/24) is not equal to previous one (0.0.0.0/
0, 0.0.0.0/0), trying to recycle old iptables rules
I0601 23:09:25.538448       1 kube.go:511] Creating the node lease for IPv4. This is the n.Spec.PodCIDRs: [10.244.1.0/24]
I0601 23:09:25.958317       1 main.go:357] Setting up masking rules
I0601 23:09:25.962264       1 main.go:408] Changing default FORWARD chain policy to ACCEPT
I0601 23:09:25.964489       1 iptables.go:290] generated 7 rules
I0601 23:09:25.967282       1 main.go:436] Wrote subnet file to /run/flannel/subnet.env
I0601 23:09:25.967338       1 main.go:440] Running backend.
I0601 23:09:25.967825       1 iptables.go:290] generated 3 rules
I0601 23:09:25.967960       1 vxlan_network.go:65] watching for new subnet leases
I0601 23:09:25.968002       1 subnet.go:159] Batch elem [0] is { lease.Event{Type:0, Lease:lease.Lease{EnableIPv4:true, EnableIPv6:false,
Subnet:ip.IP4Net{IP:0xaf40000, PrefixLen:0x18}, IPv6Subnet:ip.IP6Net{IP:(*ip.IP6)(nil), PrefixLen:0x0}, Attrs:lease.LeaseAttrs{PublicIP:0x
c0a80114, PublicIPv6:(*ip.IP6)(nil), BackendType:"vxlan", BackendData:json.RawMessage{0x7b, 0x22, 0x56, 0x4e, 0x49, 0x22, 0x3a, 0x31, 0x2c
, 0x22, 0x56, 0x74, 0x65, 0x70, 0x4d, 0x41, 0x43, 0x22, 0x3a, 0x22, 0x31, 0x65, 0x3a, 0x37, 0x64, 0x3a, 0x39, 0x36, 0x3a, 0x38, 0x37, 0x3a
, 0x38, 0x36, 0x3a, 0x65, 0x61, 0x22, 0x7d}, BackendV6Data:json.RawMessage(nil)}, Expiration:time.Date(1, time.January, 1, 0, 0, 0, 0, tim
e.UTC), Asof:0}} }
I0601 23:09:25.968282       1 subnet.go:159] Batch elem [0] is { lease.Event{Type:0, Lease:lease.Lease{EnableIPv4:true, EnableIPv6:false,
```

图 7-22　查看在指定时间之前的 Docker 容器日志

第九步，查看根据指定时间范围过滤的 Docker 容器日志，操作命令如下：

```
[root@docker ~]# docker logs --since '2024-06-02T23:30:00' --until '2024-08-13T06:00:00' a666d8908e33
```

命令运行结果如图 7-23 所示。

```
[root@docker ~]# docker logs --since '2024-06-02T23:30:00' --until '2024-08-13T06:00:00' a666d8908e33
Flag --port has been deprecated, see --secure-port instead.
I0811 14:16:16.793017       1 serving.go:331] Generated self-signed cert in-memory
I0811 14:16:18.555986       1 controllermanager.go:175] Version: v1.19.0
I0811 14:16:18.559137       1 secure_serving.go:197] Serving securely on 127.0.0.1:10257
I0811 14:16:18.559292       1 leaderelection.go:243] attempting to acquire leader lease  kube-system/kube-controller-manager...
I0811 14:16:18.560195       1 dynamic_cafile_content.go:167] Starting request-header::/etc/kubernetes/pki/front-proxy-ca.crt
I0811 14:16:18.560306       1 tlsconfig.go:240] Starting DynamicServingCertificateController
I0811 14:16:18.560708       1 dynamic_cafile_content.go:167] Starting client-ca-bundle::/etc/kubernetes/pki/ca.crt
E0811 14:16:28.560015       1 leaderelection.go:321] error retrieving resource lock kube-system/kube-controller-manager: Get "https://192.
168.1.20:6443/api/v1/namespaces/kube-system/endpoints/kube-controller-manager?timeout=10s": net/http: request canceled while waiting for c
onnection (Client.Timeout exceeded while awaiting headers)
E0811 14:16:31.699526       1 leaderelection.go:321] error retrieving resource lock kube-system/kube-controller-manager: endpoints "kube-c
ontroller-manager" is forbidden: User "system:kube-controller-manager" cannot get resource "endpoints" in API group "" in the namespace "k
ube-system": RBAC: [clusterrole.rbac.authorization.k8s.io "system:basic-user" not found, clusterrole.rbac.authorization.k8s.io "system:kub
e-controller-manager" not found, clusterrole.rbac.authorization.k8s.io "system:discovery" not found, clusterrole.rbac.authorization.k8s.io
 "system:public-info-viewer" not found]
I0811 14:16:52.639990       1 leaderelection.go:253] successfully acquired lease kube-system/kube-controller-manager
I0811 14:16:52.669108       1 event.go:291] "Event occurred" object="kube-system/kube-controller-manager" kind="Endpoints" apiVersion="v1"
 type="Normal" reason="LeaderElection" message="master_52e5e428-30ec-4469-8ec0-6814662270a8 became leader"
I0811 14:16:52.702418       1 event.go:291] "Event occurred" object="kube-system/kube-controller-manager" kind="Lease" apiVersion="coordin
ation.k8s.io/v1" type="Normal" reason="LeaderElection" message="master_52e5e428-30ec-4469-8ec0-6814662270a8 became leader"
W0811 14:16:53.180890       1 core.go:244] configure-cloud-routes is set, but no cloud provider specified. Will not configure cloud provid
er routes.
W0811 14:16:53.180922       1 controllermanager.go:541] Skipping "route"
I0811 14:16:53.184274       1 shared_informer.go:240] Waiting for caches to sync for tokens
I0811 14:16:53.278797       1 controllermanager.go:549] Started "persistentvolume-binder"
I0811 14:16:53.278992       1 pv_controller_base.go:303] Starting persistent volume controller
I0811 14:16:53.279006       1 shared_informer.go:240] Waiting for caches to sync for persistent volume
I0811 14:16:53.285476       1 shared_informer.go:247] Caches are synced for tokens
I0811 14:16:53.293695       1 controllermanager.go:549] Started "persistentvolume-expander"
I0811 14:16:53.293879       1 expand_controller.go:319] Starting expand controller
I0811 14:16:53.293890       1 shared_informer.go:240] Waiting for caches to sync for expand
I0811 14:16:53.315078       1 controllermanager.go:549] Started "pv-protection"
I0811 14:16:53.315502       1 pv_protection_controller.go:83] Starting PV protection controller
I0811 14:16:53.315520       1 shared_informer.go:240] Waiting for caches to sync for PV protection
I0811 14:16:53.335015       1 controllermanager.go:549] Started "statefulset"
I0811 14:16:53.335234       1 stateful_set.go:146] Starting stateful set controller
I0811 14:16:53.335250       1 shared_informer.go:240] Waiting for caches to sync for stateful set
```

图 7-23　查看根据指定时间范围内过滤的 Docker 容器日志

第十步，查看根据关键词过滤的 Docker 容器日志，操作命令如下：

```
[root@docker ~]# docker logs b2c37cee717c 2>&1 | grep 'error'
```

“2>&1”确保标准错误 (STDERR) 被重定向到标准输出 (STDOUT)，以便捕获错误日志。

命令运行结果如图 7-24 所示。

```
[root@docker ~]# docker logs b2c37cee717c 2>&1 | grep 'error'
W0811 14:16:18.557359       1 clientconn.go:1223] grpc: addrConn.createTransport failed to connect to {https://127.0.0.1:2379  <nil> 0 <ni
l>}. Err :connection error: desc = "transport: Error while dialing dial tcp 127.0.0.1:2379: connect: connection refused". Reconnecting...
W0811 14:31:16.695624       1 clientconn.go:1223] grpc: addrConn.createTransport failed to connect to {https://127.0.0.1:2379  <nil> 0 <ni
l>}. Err :connection error: desc = "transport: Error while dialing dial tcp 127.0.0.1:2379: connect: connection refused". Reconnecting...
W0811 14:31:16.695693       1 clientconn.go:1223] grpc: addrConn.createTransport failed to connect to {https://127.0.0.1:2379  <nil> 0 <ni
l>}. Err :connection error: desc = "transport: Error while dialing dial tcp 127.0.0.1:2379: connect: connection refused". Reconnecting...
W0811 14:31:16.725888       1 clientconn.go:1223] grpc: addrConn.createTransport failed to connect to {https://127.0.0.1:2379  <nil> 0 <ni
l>}. Err :connection error: desc = "transport: Error while dialing dial tcp 127.0.0.1:2379: connect: connection refused". Reconnecting...
W0811 14:31:16.725948       1 clientconn.go:1223] grpc: addrConn.createTransport failed to connect to {https://127.0.0.1:2379  <nil> 0 <ni
l>}. Err :connection error: desc = "transport: Error while dialing dial tcp 127.0.0.1:2379: connect: connection refused". Reconnecting...
W0811 14:31:16.725974       1 clientconn.go:1223] grpc: addrConn.createTransport failed to connect to {https://127.0.0.1:2379  <nil> 0 <ni
l>}. Err :connection error: desc = "transport: Error while dialing dial tcp 127.0.0.1:2379: connect: connection refused". Reconnecting...
W0811 14:31:16.726063       1 clientconn.go:1223] grpc: addrConn.createTransport failed to connect to {https://127.0.0.1:2379  <nil> 0 <ni
l>}. Err :connection error: desc = "transport: Error while dialing dial tcp 127.0.0.1:2379: connect: connection refused". Reconnecting...
W0811 14:31:16.726114       1 clientconn.go:1223] grpc: addrConn.createTransport failed to connect to {https://127.0.0.1:2379  <nil> 0 <ni
l>}. Err :connection error: desc = "transport: Error while dialing dial tcp 127.0.0.1:2379: connect: connection refused". Reconnecting...
W0811 14:31:16.737195       1 clientconn.go:1223] grpc: addrConn.createTransport failed to connect to {https://127.0.0.1:2379  <nil> 0 <ni
l>}. Err :connection error: desc = "transport: Error while dialing dial tcp 127.0.0.1:2379: connect: connection refused". Reconnecting...
W0811 14:31:16.737571       1 clientconn.go:1223] grpc: addrConn.createTransport failed to connect to {https://127.0.0.1:2379  <nil> 0 <ni
l>}. Err :connection error: desc = "transport: Error while dialing dial tcp 127.0.0.1:2379: connect: connection refused". Reconnecting...
W0811 14:31:16.737612       1 clientconn.go:1223] grpc: addrConn.createTransport failed to connect to {https://127.0.0.1:2379  <nil> 0 <ni
l>}. Err :connection error: desc = "transport: Error while dialing dial tcp 127.0.0.1:2379: connect: connection refused". Reconnecting...
W0811 14:31:16.851850       1 clientconn.go:1223] grpc: addrConn.createTransport failed to connect to {https://127.0.0.1:2379  <nil> 0 <ni
l>}. Err :connection error: desc = "transport: Error while dialing dial tcp 127.0.0.1:2379: connect: connection refused". Reconnecting...
W0811 14:31:16.851924       1 clientconn.go:1223] grpc: addrConn.createTransport failed to connect to {https://127.0.0.1:2379  <nil> 0 <ni
l>}. Err :connection error: desc = "transport: Error while dialing dial tcp 127.0.0.1:2379: connect: connection refused". Reconnecting...
W0811 14:31:16.851968       1 clientconn.go:1223] grpc: addrConn.createTransport failed to connect to {https://127.0.0.1:2379  <nil> 0 <ni
l>}. Err :connection error: desc = "transport: Error while dialing dial tcp 127.0.0.1:2379: connect: connection refused". Reconnecting...
W0811 14:31:16.852019       1 clientconn.go:1223] grpc: addrConn.createTransport failed to connect to {https://127.0.0.1:2379  <nil> 0 <ni
l>}. Err :connection error: desc = "transport: Error while dialing dial tcp 127.0.0.1:2379: connect: connection refused". Reconnecting...
W0811 14:31:16.852117       1 clientconn.go:1223] grpc: addrConn.createTransport failed to connect to {https://127.0.0.1:2379  <nil> 0 <ni
l>}. Err :connection error: desc = "transport: Error while dialing dial tcp 127.0.0.1:2379: connect: connection refused". Reconnecting...
W0811 14:31:16.852179       1 clientconn.go:1223] grpc: addrConn.createTransport failed to connect to {https://127.0.0.1:2379  <nil> 0 <ni
l>}. Err :connection error: desc = "transport: Error while dialing dial tcp 127.0.0.1:2379: connect: connection refused". Reconnecting...
W0811 14:31:16.852233       1 clientconn.go:1223] grpc: addrConn.createTransport failed to connect to {https://127.0.0.1:2379  <nil> 0 <ni
l>}. Err :connection error: desc = "transport: Error while dialing dial tcp 127.0.0.1:2379: connect: connection refused". Reconnecting...
W0811 14:31:16.852283       1 clientconn.go:1223] grpc: addrConn.createTransport failed to connect to {https://127.0.0.1:2379  <nil> 0 <ni
l>}. Err :connection error: desc = "transport: Error while dialing dial tcp 127.0.0.1:2379: connect: connection refused". Reconnecting...
```

图 7-24　查看根据关键词过滤的 Docker 容器日志

第十一步，将 Docker 容器日志输出到文件，以便后续离线查看和分析日志，操作命令如下：

```
[root@docker ~]# docker logs be0addc546d6 > container-be0_logs.txt
```

命令运行结果如图 7-25 所示。

```
[root@docker ~]# ls
anaconda-ks.cfg  constar.tar  exp     harbor                                 httpd  my_nginxs  nginx     sshd    systemctl  wordpress
compose_lnmp     consul       gitlab  harbor-offline-installer-v2.9.1.tgz  myh-w  mysshd1    nginx111  stress  tomcat
[root@docker ~]# docker run -d centos/httpd /bin/bash -c "while true;do echo Welcome;done"
be0addc546d6039d51ccab7caf82c157798da2f27f563ea8e0be9390c5da3937
[root@docker ~]# docker ps
CONTAINER ID   IMAGE          COMMAND                  CREATED          STATUS         PORTS    NAMES
be0addc546d6   centos/httpd   "/bin/bash -c 'while…"   10 seconds ago   Up 8 seconds   80/tcp   dazzling_pike
[root@docker ~]# docker logs be0addc546d6 > container-be0_logs.txt
^C
[root@docker ~]# ls
anaconda-ks.cfg  consul                  gitlab                                 httpd      mysshd1   sshd       tomcat
compose_lnmp     container-be0_logs.txt  harbor                                 myh-w      nginx     stress     wordpress
constar.tar      exp                     harbor-offline-installer-v2.9.1.tgz  my_nginxs  nginx111  systemctl
[root@docker ~]# cat container-be0_logs.txt
Welcome
Welcome
Welcome
Welcome
Welcome
Welcome
Welcome
Welcome
Welcome
Welcome
```

图 7-25　将 Docker 容器日志输出到文件

双创视角

天翼云在医疗行业的应用

2024 年 9 月，根据《IDC MarketScape：中国医疗云 IaaS + PaaS 2024 年厂商评估》报告显示，中国电信天翼云凭借云网融合优势、在医疗行业领先的技术、丰富的产品方案和完善的服务，入选中国医疗云 IaaS + PaaS 领导者类别。

我国医疗行业已步入数字化转型快车道，越来越多医疗机构在基础设施升级和数字化医疗建设中，倾向于选用云计算替代传统的 IT 基础设施，以进一步提升医疗服务质量与用户体验。同时，医疗影像云、医疗边缘云等场景化云解决方案以及医疗大数据和医疗 AI 开发和应用等需求，正驱动医疗云向 IaaS + PaaS 融合的方向发展，这也对云厂商医疗云基础设施服务能力提出了新的要求。

作为云服务国家队，天翼云秉承国云使命，充分发挥遍布全国的云计算资源优势和技术创新能力，在医疗卫生领域持续探索与深耕，打造领先的产品、解决方案和服务，助推医疗卫生事业向数字化、智能化加速转型。围绕医院、公共卫生、区域医疗等典型场景，天翼云构建涵盖公有云、私有云、混合云、边缘云等云服务的全场景解决方案，提供从医疗云规划设计咨询、实施服务、灾备服务等端到端的全栈服务。目前，天翼云已助力 200 余个地市级医疗云平台、8000 余家医疗机构上云，全力支撑医疗机构的云上创新发展。

项 目 小 结

本项目介绍了 Docker 安全、Cgroup 资源管理和限制机制以及 Docker 日志等内容，完成了设置容器的 CPU 使用率与 CPU 周期、限制 CPU 内核 / 内存 /Block IO 和查看 Docker 日志等操作任务，让读者学会 Docker 安全的相关知识和基本操作技能。

习 题 测 试

一、单选题

1. 与 Docker 容器相比，虚拟机的安全性是(　　)。

A. 比较低的　　B. 比较高的

C. 一样的　　D. 不确定的

2. 通过公开的镜像网站，使用未经验证或者来源不明的镜像可能包含(　　)，攻击者可以通过这种镜像进行攻击。

A. 重要数据　　B. 敏感信息

C. 端口信息　　D. 恶意代码

3. 以下描述，错误的是(　　)。

A. 未经授权的用户或者容器可能会访问其他容器或者宿主机上的敏感信息。

B. 容器之间的网络通信是安全的，攻击者不能通过监听、拦截或者篡改网络流量来进行攻击。

C. 黑客控制的恶意容器可能会消耗过多的系统资源，导致拒绝服务或者其他系统性能问题。

D. 如果缺乏对容器活动的监控和日志记录，可能导致无法追踪或分析安全事件。

二、多选题

1. Cgroup 能提供统一的接口来管理(　　)等资源。

A. CPU　　B. 内存

C. 磁盘 I/O　　D. 网络

2. Cgroup 只在容器占用资源紧缺时才起作用，Cgroup 资源配置取决于(　　)。

A. 使用镜像的数目

B. 容器运行的时间

C. 多个运行容器的 CPU 分配情况

D. 容器中进程的运行情况

3. 在实际的生产环境中，高效监控和日志管理对(　　)是至关重要的。

A. 系统的持续稳定运行　　B. 系统的故障排查

C. Docker 的网络通信　　D. 镜像仓库的管理

三、简答题

1. 简述 Cgroup 各子系统的作用。

2. 简述 Docker 日志的主要作用。

项目八

部署和管理 Docker Swarm 集群

学习目标

(1) 了解 Docker Swarm 的概念。
(2) 理解 Docker Swarm 的工作原理。
(3) 掌握配置 Docker Swarm 集群各节点的系统环境的方法。
(3) 掌握部署 Docker Swarm 集群的方法。
(4) 掌握管理 Docker Swarm 集群的方法。

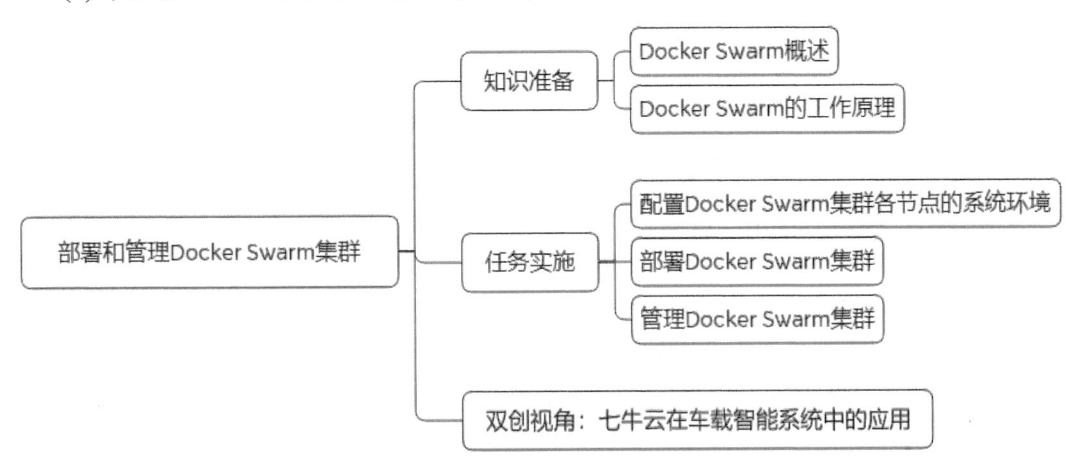

8.1　知识准备

8.1.1　Docker Swarm 概述

1. 集群

集群是由多台相互独立的主机或服务器利用通信网络组成的系统，每台主机或服务器(称为集群的节点)协同工作，作为一个整体进行管理并为用户提供服务。作为大规模数据处理和云计算等领域的基础设施，集群主要有以下特点：

(1) 高性能：集群能够跨主机或服务器提供服务，其性能远远大于单个主机或服务器。

(2) 高可用性：当集群中的某个节点发生故障时，其他节点会自动接管其工作，确保服务不会中断。

(3) 可伸缩性：当需要更多或更少的计算能力时，可以方便地在集群中添加或减少节点，实现性能的弹性伸缩。

(4) 负载均衡：任务会被均匀地分配到集群中的各个节点上，确保每个节点的工作负载都保持在合理的范围内。

(5) 数据冗余：集群中的数据会在多个节点之间进行备份，确保数据的安全性和完整性。

(6) 统一管理：集群通过统一的管理接口或工具对系统进行管理和监控。

2. Docker Swarm 的相关概念

下面介绍一些 Docker Swarm 的相关概念。

(1) 节点 (Node)：Docker Swarm 集群中的 Docker 主机。

(2) 管理节点 (Manager)：负责管理集群，包括维护集群状态、调度任务和分配资源等。默认情况下，管理节点也作为工作节点运行，但可以将其配置为仅运行集群管理任务。

(3) 工作节点 (Worker)：接收并执行 Manager 分配的任务。Worker 向 Manager 通知其执行任务的当前状态，以便 Manager 能够维持整个集群的期望状态。

(4) 服务 (Service)：在 Swarm 集群中，服务是指一组运行相同应用的容器。服务是用户与集群交互的主要根源，创建服务时需指定使用的镜像。

(5) 任务 (Task)：容器中执行的命令。任务是工作节点上调度和运行的最小单位，Manager 根据指定数量的任务副本分配任务给 Worker。

3. Docker Swarm 的基本架构

Docker Swarm 是 Docker 公司在 2014 年 12 月发布的容器编排工具，其功能与 Docker Compose 相似。Docker Compose 在单个主机或服务器上管理多个容器，而 Docker Swarm 将多台主机或服务器构成一个 Docker 集群，通过 API 管理多个主机或服务器上的 Docker 容器，并结合 Overlay 网络实现容器的调度与相互访问，更适合部署微服务。

Docker Swarm 的基本架构如图 8-1 所示。

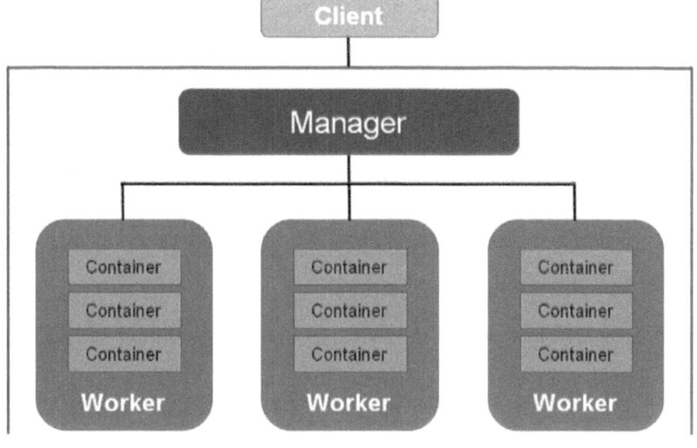

图 8-1　Docker Swarm 基本架构

Docker Swarm 具有以下功能：

(1) 容器编排：根据用户定义的规则和策略，Docker Swarm 在多个主机或服务器上自动部署和管理容器，实现容器的自动化编排。

(2) 服务发现：Docker Swarm 提供了内置的服务发现机制，允许容器之间相互通信和发现，无须手动配置 IP 地址或端口映射。

(3) 负载均衡：Docker Swam 支持负载均衡功能，可以将流量均匀地分配到多个容器之间，提高系统的性能。

(4) 弹性伸缩：Docker Swamm 能够根据实际需求动态地扩展或缩减容器的数量，以应对流量的变化和负载的增减。

(5) 滚动更新：设置更新策略，通过 Docker 服务的更新功能来实现并行更新或顺序更新。在更新过程中，Docker Swarm 会逐步替换旧的容器，直到所有容器都更新完成。

(6) 声明式服务模型：使用声明的方式定义各种所期望的服务状态，首先声明要运行的任务数，Docker Swarm 集群通过添加或删除任务来自动调整以维持所需的状态。

8.1.2　Docker Swarm 的工作原理

在 Docker Swarm 集群中，一个或多个节点被指定为管理节点 (Manager)，提高集群的高可用性，负责管理和控制集群，如维护集群的状态信息、调度容器、处理用户请求等；其他节点作为工作节点 (Worker)，负责运行服务，如接收 Manager 的任务分配，执行相应的容器操作等。Docker Swarm 使用 Raft 算法管理整个集群的状态，维护集群的状态一致性，包括服务、任务和网络的状态。Raft 算法是一种用于维护分布式系统状态一致性的复制控制协议，当集群中的管理节点发生故障时，其余的 Manager 会通过 Raft 协议选举新的领导者来维护集群的持续运行。

Docker Swarm 的工作过程如下：

(1) 初始化集群：通过运行 docker swarm init 命令指定一台主机为集群的管理节点 (Manager)。

(2) 加入 Swarm 节点：通过运行 docker swarm join 命令将其他主机作为工作节点 (Worker) 加入集群中。

(3) 部署服务：在 Manager 上，使用 docker service 命令部署服务，可以根据需求指定一个或多个副本。

(4) 调度任务：通过 Manager 上的调度器 (scheduler) 给 Worker 分配任务。

(5) 任务执行：在 Worker 上，接收并执行 Manager 分配的任务。

(6) 更新服务：通过 docker service update 命令更新服务的配置，如镜像或其他属性，确保服务能够适应不断变化的需求和环境。

(7) 扩缩和缩容：通过 docker service scale 命令调整服务副本的数量，实现扩容和缩容。

8.2 任务实施

8.2.1 配置 Docker Swarm 集群各节点的系统环境

配置 Docker Swarm 集群各节点的系统环境

1. 任务目标

掌握配置 Docker Swarm 集群各节点系统环境的方法。

2. 任务内容

(1) 克隆三台主机，并设置其 IP 地址。

(2) 使用各主机连接远程管理工具 MobaXterm。

(3) 设置各主机的主机名称。

(4) 配置各主机的主机名解析。

(5) 设置防火墙 Firewalld。

3. 完成任务所需的设备和软件

(1) 一台安装 Windows 10 操作系统的计算机。

(2) VMware Workstation、Docker。

(3) 远程管理工具 MobaXterm。

4. 任务实施步骤

第一步，利用 CentOS7-m 虚拟机克隆三台服务器，设置名称分别为 manager、worker1、worker2，Docker Swarm 集群的基本设置信息见表 8-1 所示。

表 8-1 Docker Swarm 集群的基本设置信息

主机名称	IP 地址	操作系统	主要软件	CPU 配置	内存设置
manager	192.168.1.11	CentOS 7.6	Docker CE	2 核	2 GB
worker1	192.168.1.12	CentOS 7.6	Docker CE	2 核	2 GB
worker2	192.168.1.13	CentOS 7.6	Docker CE	2 核	2 GB

以下步骤需在 manager、worker1 和 worker2 三台主机上同时运行，在此仅给出 manager 主机的运行情况，其他主机可参照进行。

第二步，开启 manager 主机，修改其 IP 地址为 192.168.1.11 并重启网络，操作命令如下：

```
[root@docker ~]# vi /etc/sysconfig/network-scripts/ifcfg-ens33
TYPE="Ethernet"
BOOTPROTO="static"
IPADDR=192.168.1.11
NETMASK=255.255.255.0
GATEWAY=192.168.1.2
NAME="ens33"
DEVICE="ens33"
ONBOOT="yes"
DNS1=114.114.114.114
DNS2=8.8.8.8
[root@docker ~]# systemctl restart network
```

第三步，通过远程管理工具 MobaXterm 连接 manager 主机，如图 8-2 所示。

图 8-2 远程管理工具 MobaXterm 连接 manager 主机

第四步，在 manager 主机上，修改其主机名为 manager，操作命令如下：

```
[root@docker ~]# hostnamectl set-hostname manager
[root@docker ~]# su
[root@manager ~]# hostname
```

命令运行结果如图 8-3 所示。

```
[root@docker ~]# hostnamectl set-hostname manager
[root@docker ~]# su
[root@manager ~]# hostname
manager
[root@manager ~]#
```

图 8-3　修改其主机名为 manager

第五步，在 manager 主机上，配置各主机的主机名解析，操作命令如下：

```
[root@manager ~]# vim /etc/hosts
127.0.0.1   localhost localhost.localdomain localhost4 localhost4.localdomain4
::1         localhost localhost.localdomain localhost6 localhost6.localdomain6
192.168.1.11 manager
192.168.1.12 worker1
192.168.1.13 worker2
```

第六步，设置防火墙 Firewalld，确保集群各节点之间的集群管理端口 TCP2377、容器网络发现端口 TCP/UDP7946、Overlay 网络通信端口 UDP4789 能够正常通信，操作命令如下：

```
[root@manager ~]# systemctl restart firewalld
[root@manager ~]# firewall-cmd --zone=public --add-port=2377/tcp --permanent
[root@manager ~]# firewall-cmd --zone=public --add-port=7946/tcp --permanent
[root@manager ~]# firewall-cmd --zone=public --add-port=7946/udp --permanent
[root@manager ~]# firewall-cmd --zone=public --add-port=4789/udp --permanent
[root@manager ~]# firewall-cmd --reload
[root@manager ~]# systemctl restart docker
```

命令运行结果如图 8-4 所示。

```
[root@manager ~]# systemctl restart firewalld
[root@manager ~]# firewall-cmd --zone=public --add-port=2377/tcp --permanent
success
[root@manager ~]# firewall-cmd --zone=public --add-port=7946/tcp --permanent
success
[root@manager ~]# firewall-cmd --zone=public --add-port=7946/udp --permanent
success
[root@manager ~]# firewall-cmd --zone=public --add-port=4789/udp --permanent
success
[root@manager ~]# firewall-cmd --reload
success
[root@manager ~]# systemctl restart docker
[root@manager ~]#
```

图 8-4　设置防火墙 Firewalld

第七步，根据需要为 manager 主机创建快照，然后保存此时的状态。

8.2.2　部署 Docker Swarm 集群

部署 Docker Swarm 集群

1. 任务目标

掌握部署 Docker Swarm 集群的方法。

2. 任务内容

(1) 创建 Docker Swarm 集群。

(2) 添加工作节点到 Docker Swarm 集群。

(3) 查看 Docker Swarm 集群中各节点的状态。

3. 完成任务所需的设备和软件

(1) 一台安装 Windows 10 操作系统的计算机。

(2) VMware Workstation、Docker。

(3) 远程管理工具 MobaXterm。

4. 任务实施步骤

第一步，创建 Docker Swarm 集群，操作命令如下：

```
[root@manager ~]# docker swarm init --advertise-addr 192.168.1.11
```

--advertise-addr：用于指定 Docker Swarm 集群中管理节点的 IP 地址。

命令运行结果如图 8-5 所示。

```
[root@manager ~]# docker swarm init --advertise-addr 192.168.1.11
Swarm initialized: current node (51408ntosibx9r801ujewfski) is now a manager.

To add a worker to this swarm, run the following command:

    docker swarm join --token SWMTKN-1-6bcsufe95qp9idqc2d3l4ck3aph6jxng92arb3krknmivbvd9i-39kcqlw3w1aghrbsa9h9vhyf3 192.168.1.11:2377

To add a manager to this swarm, run 'docker swarm join-token manager' and follow the instructions.

[root@manager ~]#
```

图 8-5　创建 Docker Swarm 集群

从命令运行结果的输出信息可以看到：

(1) 集群已经创建成功，当前节点是一个管理者。

(2) 当工作节点加入该集群时，运行以下命令：

```
docker swarm join --token SWMTKN-1-6bcsufe95qp9idqc2d3l4ck3aph6jxng92arb3krknmivbvd9i-39kcqlw3w1aghrbsa9h9vhyf3 192.168.1.11:2377
```

注意：如果看不到这个信息，可以通过运行命令 docker swarm join-token worker 得到。

(3) 管理节点加入该集群时，运行以下命令：

```
docker swarm join-token manager
```

命令运行结果为：管理节点加入集群要运行的命令，如图 8-6 所示。

```
[root@manager ~]# docker swarm join-token manager
To add a manager to this swarm, run the following command:

    docker swarm join --token SWMTKN-1-6bcsufe95qp9idqc2d3l4ck3aph6jxng92arb3krknmivbvd9i-68gn2xfpxgtd8jxu9dff78tyc 192.168.1.11:2377

[root@manager ~]#
```

图 8-6　管理节点加入集群要运行的命令

第二步，将工作节点 worker1 加入 Docker Swarm 集群中，操作命令如下：

```
[root@worker1 ~]# docker swarm join --token SWMTKN-1-6bcsufe95qp9idqc2d3l4ck3aph6jxng92arb3krknmivbvd9i-39kcqlw3w1aghrbsa9h9vhyf3 192.168.1.11:2377
```

命令运行结果如图 8-7 所示。

```
[root@worker1 ~]# docker swarm join --token SWMTKN-1-6bcsufe95qp9idqc2d3l4ck3aph6jxng92arb3krknmivbvd9i-39kcqlw3w1aghrbsa9h9vhyf3 192.168.
1.11:2377
This node joined a swarm as a worker.
[root@worker1 ~]#
```

图 8-7　将工作节点 worker1 加入 Docker Swarm 集群中

第三步，将工作节点 worker2 加入 Docker Swarm 集群中，操作命令如下：

```
[root@worker2 ~]# docker swarm join --token SWMTKN-1-6bcsufe95qp9idqc2d3l4ck3aph6jxng92arb3kr
knmivbvd9i-39kcqlw3w1aghrbsa9h9vhyf3 192.168.1.11:2377
```

命令运行结果如图 8-8 所示。

```
[root@worker2 ~]# docker swarm join --token SWMTKN-1-6bcsufe95qp9idqc2d3l4ck3aph6jxng92arb3krknmivbvd9i-39kcqlw3w1aghrbsa9h9vhyf3 192.168.
1.11:2377
This node joined a swarm as a worker.
[root@worker2 ~]#
```

图 8-8　将工作节点 worker2 加入 Docker Swarm 集群中

第四步，查看集群信息，操作命令如下：

```
[root@manager ~]# docker info
```

命令运行结果如图 8-9 所示。

```
[root@manager ~]# docker info
Client:
 Context:    default
 Debug Mode: false
 Plugins:
  app: Docker App (Docker Inc., v0.9.1-beta3)
  buildx: Docker Buildx (Docker Inc., v0.8.1-docker)
  scan: Docker Scan (Docker Inc., v0.17.0)

Server:
 Containers: 0
  Running: 0
  Paused: 0
  Stopped: 0
 Images: 3
 Server Version: 20.10.14
 Storage Driver: overlay2
  Backing Filesystem: xfs
  Supports d_type: true
  Native Overlay Diff: true
  userxattr: false
 Logging Driver: json-file
 Cgroup Driver: cgroupfs
 Cgroup Version: 1
 Plugins:
  Volume: local
  Network: bridge host ipvlan macvlan null overlay
  Log: awslogs fluentd gcplogs gelf journald json-file local logentries splunk syslog
 Swarm: active
  NodeID: 51408ntosibx9r801ujewfski
  Is Manager: true
  ClusterID: jrz3ecu05lq7r2dl0lp7q21ow
  Managers: 1
  Nodes: 3
  Default Address Pool: 10.0.0.0/8
  SubnetSize: 24
  Data Path Port: 4789
  Orchestration:
   Task History Retention Limit: 5
```

图 8-9　查看集群信息

第五步，查看集群中所有节点的状态信息，操作命令如下：

```
[root@manager ~]# docker node ls
```

命令运行结果如图 8-10 所示。

```
[root@manager ~]# docker node ls
ID                            HOSTNAME   STATUS    AVAILABILITY   MANAGER STATUS   ENGINE VERSION
51408ntosibx9r801ujewfski *   manager    Ready     Active         Leader           20.10.14
hrzkn6f33xmhpcpda5ehswucz     worker1    Ready     Active                          20.10.14
byddjw817exz7mx0jm9re6h40     worker2    Ready     Active                          20.10.14
[root@manager ~]#
```

图 8-10　查看集群中各节点的状态信息

第六步，查看管理节点的详细信息，操作命令如下：

[root@manager ~]# docker node inspect manager

命令运行结果如图 8-11 所示。

```
[root@manager ~]# docker node inspect manager
[
    {
        "ID": "51408ntosibx9r801ujewfski",
        "Version": {
            "Index": 9
        },
        "CreatedAt": "2024-08-15T04:23:43.940813679Z",
        "UpdatedAt": "2024-08-15T04:23:44.476469269Z",
        "Spec": {
            "Labels": {},
            "Role": "manager",
            "Availability": "active"
        },
        "Description": {
            "Hostname": "manager",
            "Platform": {
                "Architecture": "x86_64",
                "OS": "linux"
            },
            "Resources": {
                "NanoCPUs": 2000000000,
                "MemoryBytes": 1907970048
            },
            "Engine": {
                "EngineVersion": "20.10.14",
                "Plugins": [
                    {
                        "Type": "Log",
                        "Name": "awslogs"
                    },
                    {
                        "Type": "Log",
                        "Name": "fluentd"
                    },
                    {
                        "Type": "Log",
                        "Name": "gcplogs"
                    },
```

图 8-11　查看管理节点的详细信息

8.2.3　管理 Docker Swarm 集群

管理 Docker Swarm 集群

1. 任务目标

掌握管理 Docker Swarm 集群的方法。

2. 任务内容

(1) 管理 Docker Swarm 集群的服务。

(2) 管理 Docker Swarm 集群的网络。

(3) 管理 Docker Swarm 集群的数据卷。

3. 完成任务所需的设备和软件

(1) 一台安装 Windows 10 操作系统的计算机。

(2) VMware Workstation、Docker。

(3) 远程管理工具 MobaXterm。

4. 任务实施步骤

第一步，利用 Docker 镜像 centos/httpd 创建一个名为 cht 的服务，指定服务副本数为 2，操作命令如下：

```
[root@manager ~]# docker service create --replicas 2 --name cht centos/httpd
```

命令运行结果如图 8-12 所示。

```
[root@manager ~]# docker service create --replicas 2 --name cht centos/httpd
0mqjpfz5243siac9kg0eyd945
overall progress: 2 out of 2 tasks
1/2: running   [==================================================>]
2/2: running   [==================================================>]
verify: Service converged
[root@manager ~]#
```

图 8-12　创建一个名为 cht 的服务并指定服务副本数为 2

第二步，查看已经部署启动的全部服务，操作命令如下：

```
[root@manager ~]# docker service ls
```

命令运行结果如图 8-13 所示。

```
[root@manager ~]# docker service ls
ID             NAME      MODE         REPLICAS   IMAGE                  PORTS
0mqjpfz5243s   cht       replicated   2/2        centos/httpd:latest
[root@manager ~]#
```

图 8-13　查看已经部署启动的全部服务

输出信息说明服务的基本状态和配置情况：

ID：服务的唯一标识符。

NAME：服务的名称。

MODE：服务的部署模式，包括 replicated(指定数量的副本) 和 global(每个节点一个副本)。

REPLICAS：当前运行的副本数量和服务期望的副本数量。

IMAGE：服务使用的镜像名称。

PORTS：服务暴露的端口信息。

第三步，查看指定服务的详细信息，操作命令如下：

```
[root@manager ~]# docker service ps cht
```

命令运行结果如图 8-14 所示。

```
[root@manager ~]# docker service ps cht
ID             NAME    IMAGE                 NODE      DESIRED STATE   CURRENT STATE            ERROR     PORTS
gtz0rj0bp6yh   cht.1   centos/httpd:latest   manager   Running         Running 44 minutes ago
4e631ujxwrhq   cht.2   centos/httpd:latest   worker1   Running         Running 44 minutes ago
[root@manager ~]#
```

图 8-14　查看指定服务的详细信息

第四步，在管理节点上查看当前启动的容器，操作命令如下：

```
[root@manager ~]# docker ps
```

命令运行结果如图 8-15 所示。

```
[root@manager ~]# docker ps
CONTAINER ID   IMAGE                 COMMAND            CREATED          STATUS          PORTS     NAMES
1cf51dce6b76   centos/httpd:latest   "/run-httpd.sh"    50 minutes ago   Up 49 minutes   80/tcp    cht.1.gtz0rj0bp6yhpy7k703yzsr1g
[root@manager ~]#
```

图 8-15　在管理节点上查看当前启动的容器

第五步，在工作节点 worker1 上查看当前启动的容器，操作命令如下：

[root@worker1 ~]# docker ps

命令运行结果如图 8-16 所示。

```
[root@worker1 ~]# docker ps
CONTAINER ID   IMAGE                 COMMAND           CREATED          STATUS          PORTS    NAMES
23af003b5c94   centos/httpd:latest   "/run-httpd.sh"   49 minutes ago   Up 49 minutes   80/tcp   cht.2.4e631ujxwrhqjt77kpmt47tkl
[root@worker1 ~]#
```

图 8-16　在工作节点 Worker1 上查看当前启动的容器

第六步，以易于阅读的方式，显示指定服务的详细信息，操作命令如下：

[root@manager ~]# docker service inspect --pretty cht

命令运行结果如图 8-17 所示。

```
[root@manager ~]# docker service inspect --pretty cht

ID:             0mqjpfz5243siac9kg0eyd945
Name:           cht
Service Mode:   Replicated
 Replicas:      2
Placement:
UpdateConfig:
 Parallelism:   1
 On failure:    pause
 Monitoring Period: 5s
 Max failure ratio: 0
 Update order:      stop-first
RollbackConfig:
 Parallelism:   1
 On failure:    pause
 Monitoring Period: 5s
 Max failure ratio: 0
 Rollback order:    stop-first
ContainerSpec:
 Image:         centos/httpd:latest@sha256:26c6674463ff3b8529874b17f8bb55d21a0dcf86e025eafb3c9eeee15ee4f369
 Init:          false
Resources:
Endpoint Mode:  vip
```

图 8-17　以易于阅读的方式显示指定服务的详细信息

第七步，将 cht 服务的 2 个副本扩容到 3 个副本，并查看扩容后的服务信息，操作命令如下：

[root@manager ~]# docker service scale cht=3

[root@manager ~]# docker service ps cht

命令运行结果如图 8-18 所示。

```
[root@manager ~]# docker service scale cht=3
cht scaled to 3
overall progress: 3 out of 3 tasks
1/3: running   [==================================================>]
2/3: running   [==================================================>]
3/3: running   [==================================================>]
verify: Service converged
[root@manager ~]# docker service ps cht
ID             NAME    IMAGE                 NODE      DESIRED STATE   CURRENT STATE           ERROR     PORTS
gtz0rj0bp6yh   cht.1   centos/httpd:latest   manager   Running         Running 2 hours ago
4e631ujxwrhq   cht.2   centos/httpd:latest   worker1   Running         Running 2 hours ago
lyp8rzfi6kkk   cht.3   centos/httpd:latest   worker2   Running         Running 5 minutes ago
[root@manager ~]#
```

图 8-18　将 cht 服务扩容到 3 个副本并查看服务信息

第八步，将 cht 服务的 3 个副本缩容到 1 个副本，并查看缩容后的服务信息，操作命令如下：

[root@manager ~]# docker service scale cht=3

[root@manager ~]# docker service ps cht

命令运行结果如图 8-19 所示。

```
[root@manager ~]# docker service scale cht=1
cht scaled to 1
overall progress: 1 out of 1 tasks
1/1: running   [==================================================>]
verify: Service converged
[root@manager ~]# docker service ps cht
ID             NAME      IMAGE                  NODE      DESIRED STATE   CURRENT STATE           ERROR     PORTS
gtz0rj0bp6yh   cht.1     centos/httpd:latest    manager   Running         Running 2 hours ago
[root@manager ~]#
```

图 8-19　将 cht 服务缩容到 1 个副本并查看服务信息

第九步，删除集群中所有的 cht 服务，操作命令如下：

```
[root@manager ~]# docker service rm cht
[root@manager ~]# docker service ps cht
```

命令运行结果如图 8-20 所示。

```
[root@manager ~]# docker service rm cht
cht
[root@manager ~]# docker service ps cht
no such service: cht
[root@manager ~]#
```

图 8-20　删除集群中所有的 cht 服务

第十步，Docker Swarm 集群将服务仅部署在工作节点上，操作步骤如下：

(1) 查看集群中所有工作节点的状态信息，操作命令如下：

```
[root@manager ~]# docker node ls
[root@manager ~]# docker node inspect worker1
[root@manager ~]# docker node inspect worker2
```

命令运行结如图 8-21 所示。

```
[root@manager ~]# docker node ls
ID                            HOSTNAME   STATUS    AVAILABILITY   MANAGER STATUS   ENGINE VERSION
51408ntosibx9r801ujewfski *   manager    Ready     Active         Leader           20.10.14
hrzkn6f33xmhpcpda5ehswucz     worker1    Ready     Active                          20.10.14
byddjw817exz7mx0jm9re6h40     worker2    Ready     Active                          20.10.14
[root@manager ~]# docker node inspect worker1
[
    {
        "ID": "hrzkn6f33xmhpcpda5ehswucz",
        "Version": {
            "Index": 341
        },
        "CreatedAt": "2024-08-15T05:05:03.312687525Z",
        "UpdatedAt": "2024-08-15T23:55:19.824319206Z",
        "Spec": {
            "Labels": {},
            "Role": "worker",
            "Availability": "active"
        },
        "Description": {
            "Hostname": "worker1",
            "Platform": {
                "Architecture": "x86_64",
                "OS": "linux"
            },
            "Resources": {
                "NanoCPUs": 2000000000,
                "MemoryBytes": 1907970048
            },
            "Engine": {
                "EngineVersion": "20.10.14",
                "Plugins": [
                    {
                        "Type": "Log",
                        "Name": "awslogs"
                    },
                    {
```

图 8-21　查看集群中所有工作节点的状态信息

(2) 为工作节点 Worker1 添加标签，并查看添加情况，操作命令如下：

[root@manager ~]# docker node update --label-add role=worker hrzkn6f33xmhpcpda5ehswucz(工作节点 worker1 的 ID)

[root@manager ~]# docker node inspect hrzkn6f33xmhpcpda5ehswucz

命令运行结果如图 8-22 所示。

```
[root@manager ~]# docker node update --label-add role=worker hrzkn6f33xmhpcpda5ehswucz
hrzkn6f33xmhpcpda5ehswucz
[root@manager ~]# docker node inspect hrzkn6f33xmhpcpda5ehswucz
[
    {
        "ID": "hrzkn6f33xmhpcpda5ehswucz",
        "Version": {
            "Index": 341
        },
        "CreatedAt": "2024-08-15T05:05:03.312687525Z",
        "UpdatedAt": "2024-08-15T23:55:19.824319206Z",
        "Spec": {
            "Labels": {
                "role": "worker"
            },
            "Role": "worker",
            "Availability": "active"
        },
        "Description": {
            "Hostname": "worker1",
            "Platform": {
                "Architecture": "x86_64",
                "OS": "linux"
            },
            "Resources": {
                "NanoCPUs": 2000000000,
                "MemoryBytes": 1907970048
            },
            "Engine": {
                "EngineVersion": "20.10.14",
                "Plugins": [
                    {
                        "Type": "Log",
                        "Name": "awslogs"
                    },
                    {
                        "Type": "Log",
```

图 8-22　为工作节点 Worker1 添加标签

(3) 为工作节点 Worker2 添加标签，并查看添加情况，操作命令如下：

[root@manager ~]# docker node update --label-add role=worker byddjw817exz7mx0jm9re6h40

[root@manager ~]# docker node inspect byddjw817exz7mx0jm9re6h40

命令运行结果如图 8-23 所示。

```
[root@manager ~]# docker node update --label-add role=worker byddjw817exz7mx0jm9re6h40
byddjw817exz7mx0jm9re6h40
[root@manager ~]# docker node inspect byddjw817exz7mx0jm9re6h40
[
    {
        "ID": "byddjw817exz7mx0jm9re6h40",
        "Version": {
            "Index": 342
        },
        "CreatedAt": "2024-08-15T05:08:21.389241484Z",
        "UpdatedAt": "2024-08-15T23:55:59.121704348Z",
        "Spec": {
            "Labels": {
                "role": "worker"
            },
            "Role": "worker",
            "Availability": "active"
        },
        "Description": {
            "Hostname": "worker2",
            "Platform": {
                "Architecture": "x86_64",
                "OS": "linux"
            },
            "Resources": {
```

```
        "NanoCPUs": 2000000000,
        "MemoryBytes": 1907970048
    },
    "Engine": {
        "EngineVersion": "20.10.14",
        "Plugins": [
            {
                "Type": "Log",
                "Name": "awslogs"
            },
            {
                "Type": "Log",
```

图 8-23　为工作节点 Worker2 添加标签

(4) 创建一个名为 chtt 的服务，指定服务副本数为 2，并查看服务的详细信息，操作命令如下：

```
[root@manager ~]# docker service create --replicas 2 --constraint 'node.labels.role==worker' --name chtt centos/httpd
[root@manager ~]# docker service ls
[root@manager ~]# docker service ps chtt
```

--constraint：仅在标签与表达式匹配的节点上部署服务。

命令运行结果如图 8-24 所示。

```
[root@manager ~]# docker service create --replicas 2 --constraint 'node.labels.role==worker' --name chtt centos/httpd
qv7jqeabx9pf3xwkow771or7v
overall progress: 2 out of 2 tasks
1/2: running   [==================================================>]
2/2: running   [==================================================>]
verify: Service converged
[root@manager ~]#  docker service ls
ID             NAME      MODE         REPLICAS   IMAGE                 PORTS
qv7jqeabx9pf   chtt      replicated   2/2        centos/httpd:latest
[root@manager ~]#  docker service ps chtt
ID             NAME      IMAGE                 NODE      DESIRED STATE   CURRENT STATE            ERROR     PORTS
qm9l7mclvmzy   chtt.1    centos/httpd:latest   worker1   Running         Running 41 seconds ago
shzk2mivo3i8   chtt.2    centos/httpd:latest   worker2   Running         Running 41 seconds ago
[root@manager ~]#
```

图 8-24　创建服务 chtt 并查看服务的详细信息

可以看出，此时服务仅部署在了工作节点上。

第十一步，在管理节点上创建一个 Overlay 网络，命名为 my_network，操作命令如下：

```
[root@manager ~]# docker network create --driver overlay my_network
[root@manager ~]# docker network ls
```

命令运行结果如图 8-25 所示。

```
[root@manager ~]# docker network create --driver overlay my_network
klwg1n1wuo2ufc5nrvtffm10c
[root@manager ~]# docker network ls
NETWORK ID     NAME         DRIVER    SCOPE
d2bcbd35d760   bridge       bridge    local
5d2228dc2005   host         host      local
klwg1n1wuo2u   my_network   overlay   swarm
4d9e4a88afee   none         null      local
[root@manager ~]#
```

图 8-25　在管理节点上创建一个 Overlay 网络

第十二步，创建一个名为 cehtt 的服务，指定服务副本数为 3，使得处于同一个 Overlay 网络中的所有服务之间可以相互通信，操作命令如下：

```
[root@manager ~]# docker service create --replicas 3 --network my_network --name cehtt centos/httpd
```

命令运行结果如图 8-26 所示。

```
[root@manager ~]# docker service create --replicas 3 --network my_network --name cehtt centos/httpd
slemblcyu9kjccnp7j5l14vb1
overall progress: 3 out of 3 tasks
1/3: running   [==================================================>]
2/3: running   [==================================================>]
3/3: running   [==================================================>]
verify: Service converged
[root@manager ~]#
```

图 8-26 创建一个服务 cehtt

第十三步，创建数据卷，操作命令如下：

[root@manager ~]# docker volume create mydata

[root@manager ~]# docker volume ls

命令运行结果如图 8-27 所示。

```
[root@manager ~]# docker volume create mydata
mydata
[root@manager ~]# docker volume ls
DRIVER    VOLUME NAME
local     mydata
[root@manager ~]#
```

图 8-27 创建数据卷

第十四步，创建服务应用数据卷，并查看服务信息，操作命令如下：

[root@manager ~]# docker service create --mount type=volume,src=mydata,dst=/usr/share/httpd --replicas 2 --name my_data_1 centos/httpd

[root@manager ~]# docker service ls

[root@manager ~]# docker service ps my_data_1

命令运行结果如图 8-28 所示。

```
[root@manager ~]# docker service create --mount type=volume,src=mydata,dst=/usr/share/httpd --replicas 2 --name my_data_1 centos/httpd
r5diastoe2apdhb95i2wz4u5r
overall progress: 2 out of 2 tasks
1/2: running   [==================================================>]
2/2: running   [==================================================>]
verify: Service converged
[root@manager ~]# docker service ls
ID             NAME        MODE         REPLICAS   IMAGE                 PORTS
r5diastoe2ap   my_data_1   replicated   2/2        centos/httpd:latest
[root@manager ~]# docker service ps my_data_1
ID             NAME          IMAGE                 NODE      DESIRED STATE   CURRENT STATE              ERROR     PORTS
pz68j3b47n74   my_data_1.1   centos/httpd:latest   manager   Running         Running about a minute ago
lvojdxxdwjfo   my_data_1.2   centos/httpd:latest   worker2   Running         Running about a minute ago
```

图 8-28 创建服务应用数据卷

第十五步，查看数据卷的详细信息，操作命令如下：

[root@manager ~]# docker volume inspect mydata

命令运行结果如图 8-29 所示。

```
[root@manager ~]# docker volume inspect mydata
[
    {
        "CreatedAt": "2024-08-16T13:34:09+08:00",
        "Driver": "local",
        "Labels": {},
        "Mountpoint": "/var/lib/docker/volumes/mydata/_data",
        "Name": "mydata",
        "Options": {},
        "Scope": "local"
    }
]
```

图 8-29 查看数据卷的详细信息

第十六步，在管理节点上检验数据是否同步，操作命令如下：

```
[root@manager ~]# cd /var/lib/docker/volumes/mydata/_data
[root@manager _data]# touch file1 file2
[root@manager _data]# ls
[root@manager _data]# cd
[root@manager ~]# docker ps
[root@manager ~]# docker exec -it a01274318dd2 bash
[root@a01274318dd2 /]# ls /usr/share/httpd
```

命令运行结果如图 8-30 所示。

```
[root@manager ~]# cd /var/lib/docker/volumes/mydata/_data
[root@manager _data]# touch file1 file2
[root@manager _data]# ls
error  file1  file2  icons  noindex
[root@manager _data]# cd
[root@manager ~]# docker ps
CONTAINER ID   IMAGE                  COMMAND             CREATED         STATUS         PORTS     NAMES
a01274318dd2   centos/httpd:latest   "/run-httpd.sh"   5 minutes ago   Up 5 minutes   80/tcp    my_data_1.1.pz68j3b47n74ae97n8tm8w3vd
[root@manager ~]# docker exec -it a01274318dd2 bash
[root@a01274318dd2 /]# ls /usr/share/httpd
error  file1  file2  icons  noindex
[root@a01274318dd2 /]#
```

图 8-30　在管理节点上检验数据是否同步

第十七步，在工作节点 Worker2 上检验数据是否同步，操作命令如下：

```
[root@worker2 ~]# docker volume inspect mydata
[root@worker2 ~]# cd /var/lib/docker/volumes/mydata/_data
[root@worker2 _data]# touch file3 file4
[root@worker2 _data]# ls
[root@worker2 _data]# cd
[root@worker2 ~]# docker ps
[root@worker2 ~]# docker exec -it 4fd2a36693bc bash
[root@4fd2a36693bc /]# ls /usr/share/httpd
```

命令运行结果如图 8-31 所示。

```
[root@worker2 ~]# docker volume inspect mydata
[
    {
        "CreatedAt": "2024-08-16T13:34:09+08:00",
        "Driver": "local",
        "Labels": null,
        "Mountpoint": "/var/lib/docker/volumes/mydata/_data",
        "Name": "mydata",
        "Options": null,
        "Scope": "local"
    }
]
[root@worker2 ~]# cd /var/lib/docker/volumes/mydata/_data
[root@worker2 _data]# touch file3 file4
[root@worker2 _data]# ls
error  file3  file4  icons  noindex
[root@worker2 _data]# cd
[root@worker2 ~]# docker ps
CONTAINER ID   IMAGE                  COMMAND             CREATED          STATUS          PORTS     NAMES
4fd2a36693bc   centos/httpd:latest   "/run-httpd.sh"   31 minutes ago   Up 31 minutes   80/tcp    my_data_1.2.lvojdxxdwjfo8xlbv9li5z041
[root@worker2 ~]# docker exec -it 4fd2a36693bc bash
[root@4fd2a36693bc /]# ls /usr/share/httpd
error  file3  file4  icons  noindex
[root@4fd2a36693bc /]#
```

图 8-31　在工作节点 Worker2 上检验数据是否同步

第十八步，在管理节点上删除服务之后，查看节点中的数据是否存在，操作命令如下：

```
[root@a01274318dd2 /]# exit
```

```
[root@manager ~]# docker service ls
[root@manager ~]# docker service rm my_data_1
[root@manager ~]# docker ps
[root@manager ~]# cd /var/lib/docker/volumes/mydata/_data
[root@manager _data]# ls
```

命令运行结果如图 8-32 所示。

```
[root@a01274318dd2 /]# exit
exit
[root@manager ~]# docker service ls
ID            NAME       MODE        REPLICAS   IMAGE                 PORTS
r5diastoe2ap  my_data_1  replicated  2/2        centos/httpd:latest
[root@manager ~]# docker service rm my_data_1
my_data_1
[root@manager ~]# docker ps
CONTAINER ID   IMAGE    COMMAND   CREATED   STATUS   PORTS    NAMES
[root@manager ~]#  cd /var/lib/docker/volumes/mydata/_data
[root@manager _data]# ls
error  file1  file2  icons  noindex
[root@manager _data]#
```

图 8-32　删除管理节点上的服务并查看节点中的数据

第十九步，删除服务之后，在工作节点 Worker2 上查看数据是否存在，操作命令如下：

```
[root@4fd2a36693bc /]# exit
[root@worker2 ~]# docker ps
[root@worker2 ~]# cd /var/lib/docker/volumes/mydata/_data
[root@worker2 _data]# ls
```

命令运行结果如图 8-33 所示。

```
[root@4fd2a36693bc /]# exit
exit
[root@worker2 ~]# docker ps
CONTAINER ID   IMAGE    COMMAND   CREATED   STATUS   PORTS    NAMES
[root@worker2 ~]# cd /var/lib/docker/volumes/mydata/_data
[root@worker2 _data]# ls
error  file3  file4  icons  noindex
[root@worker2 _data]#
```

图 8-33　删除服务后在工作节点 Worker2 上查看数据是否存在

可见，在部署了服务的两个节点上，实现了数据的可持久化。

双创视角

七牛云在车载智能系统中的应用

七牛云携手合作伙伴基于车载摄像头、AI 智能盒子、直播与实时互动等技术打造可视化安全出行、车联网等场景化解决方案，帮助车企实现数字化转型和升级，加快产品和服务创新。

面向乘用车和商用车，七牛云为驾驶员行为监管、货物搬运检测、行车路况及停车安全监控提供可视化视频管理业务，包括视频连接上云、云上按需录制、按需截图等功能。其优点包括：支持现有设备无须改造即可上云；提供标准、低频、归档三种存储类型，满足客户对存储性能、成本的不同诉求；集视频截图、录制、检索、对讲、PTZ 控制等功能于一体。

基于七牛云 LiveNet 实时流网络技术，七牛云视频监控 QVS 在云端收流后，跨地域跨运营商分发加速，可体验极速视频服务。警告关联视频上传七牛云对象存储 Kodo 后，结合断点续传及异地容灾能力，确保关键视频在云上永久存储，可靠性高。基于七牛云丰富的 AI 算法模型与海量训练结果，车载智能系统提供防止疲劳驾驶、违禁品运输、保障行车安全等场景化 AI 功能，大幅提升用车安全、满足行业合规要求；在服务端实现全天候质量监测，实时感知各类情况，保障音视频互动体验。

项 目 小 结

本项目介绍了 Docker Swarm 的基本知识及其工作原理等内容，完成了配置 Docker Swarm 集群各节点的系统环境、部署和管理 Docker Swarm 集群等操作任务，可以让读者对 Docker Swarm 集群有一定认识。

习 题 测 试

一、单选题

1. (　　) 是由多台相互独立的主机或服务器利用通信网络组成的系统，每台主机或服务器协同工作，作为一个整体进行管理并为用户提供服务。

A. 计算机　　B. 虚拟机

C. 仓库　　D. 集群

2. (　　) 是指当集群中的某个节点发生故障时，其他节点会自动接替其工作，确保服务不会中断。

A. 高性能　　B. 高可用性

C. 可伸缩性　　D. 负载均衡

3. 集群通过统一的 (　　) 或工具对系统进行管理和监控。

A. 工作节点　　B. 管理节点

C. 管理接口　　D. 网络接口

二、多选题

1. Docker Swarm 集群中运行的多个节点被分为 (　　)。

A. 存储节点　　B. 网络节点

C. 管理节点　　D. 工作节点

2. 管理节点负责管理集群，包括 (　　) 等。

A. 维护集群状态　　B. 调度任务

C. 分配资源　　D. 存储数据

3. 工作节点 (Worker) 负责运行服务，如 (　　) 等。

A. 接收用户请求　　B. 接收 Manager 的任务分配

C. 执行创建镜像的操作　　D. 执行相应的容器操作

三、简答题

1. 简述 Docker Swarm 具有哪些功能。

2. 简述 Docker Swarm 的工作过程。

项目九

部署和管理 Kubernetes 集群

学习目标

(1) 了解 Kubernetes 的概念。
(2) 理解 Kubernetes 的体系架构。
(3) 理解 Kubernetes 的相关概念。
(4) 了解 Kubernetes 集群的管理。
(5) 掌握配置 Kubernetes 集群各节点系统环境的方法。
(6) 掌握部署 Kubernetes 集群的方法。
(7) 掌握 Kubectl 的基本操作。
(8) 掌握通过 YAML 文件创建 Pod 的方法。
(9) 掌握通过标签调度 Pod 的方法。
(10) 掌握通过 YAML 文件创建 Deployment 的方法。
(11) 掌握多容器共享 Volume 的方法。

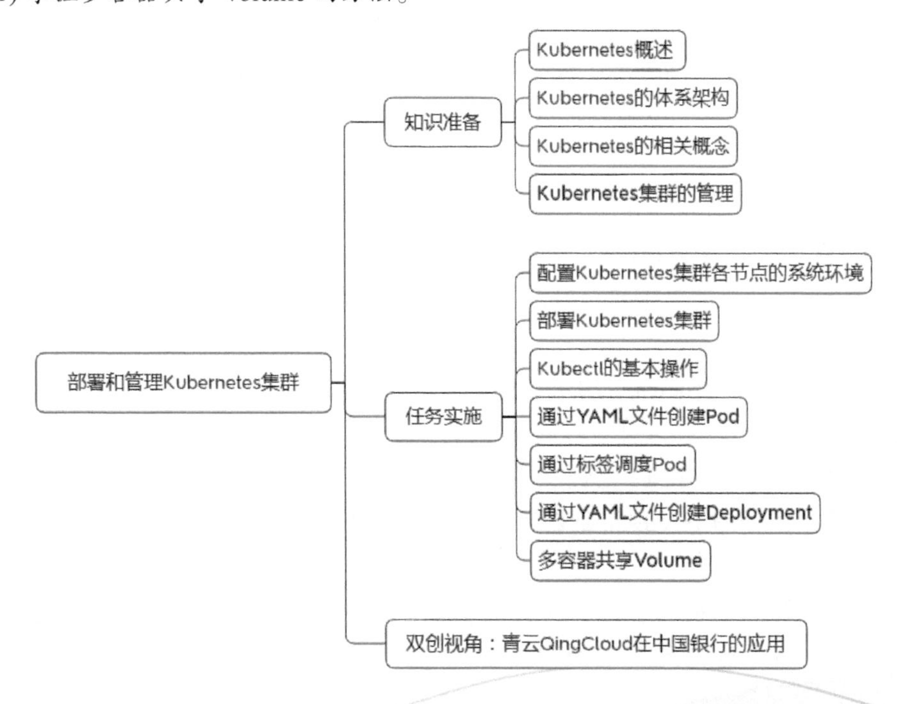

9.1 知识准备

9.1.1 Kubernetes 概述

Kubernetes 在希腊语中的意思是“舵手”，由于字母 K 与 s 之间有 8 个字母，所以人们简称其为 K8S，它是由 Google 主导开发的容器编排系统，用于管理云平台中多个主机上的容器化应用。Kubernetes 基于 Go 语言开发，前身是 Google 公司开发的 Borg 系统。Borg 系统在 Google 内部已经应用了十几年，Google 公司将 Borg 系统完善后贡献给了开源社区，并将其重新命名为 Kubernetes。

Kubernetes 系统支持用户通过模板定义服务配置，用户提交配置信息后，系统会自动完成对应用容器的创建、部署、发布、伸缩、更新等操作。系统发布后吸引了众多知名互联网公司与容器爱好者的关注，是目前容器集群管理系统中优秀的开源项目之一。

1. Kubernetes 的主要功能

(1) 容器编排：Kubernetes 管理集群中容器的调度，维持应用程序所需的状态。

(2) 自我修复：在节点产生故障时，在预期副本数量不变的情况下，终止健康检查失败的容器并部署新的容器，保证服务不会中断。

(3) 存储部署：Kubernetes 挂载外部存储系统，将这些存储作为集群资源的一部分，增强存储的灵活性。

(4) 弹性伸缩：Kubernetes 可以使用命令或基于 CPU 使用情况快速启动扩容和缩容应用程序，保证在高峰期的高可用性和业务低谷期回收资源。

(5) 资源监控：工作节点集成了 Advisor 资源收集工具，可以快速实现对集群资源的监控。

(6) 认证授权：限制用户是否有权限使用 API 进行操作，精细化权限分配。

(7) 密钥管理：Kubernetes 允许存储和管理敏感信息，如密码、OAuth 令牌和 SSH 密钥。用户可以部署或更新机密和应用程序配置，而无须重建容器镜像，不会在堆栈配置中暴露机密。

(8) 回滚更新：Kubernetes 采用滚动更新策略更新应用，一次更新一个 Pod。当更新过程中出现问题时，Kubernetes 会进行回滚更新，保证升级业务不受影响。

(9) 服务发现：Kubernetes 允许一个应用程序找到和链接到 Kubernetes 集群中的其他服务，通常采用的服务发现机制会为同一个 pod 里的容器自动设置环境变量、服务的 DNS 记录等。

(10) 负载均衡：将网络请求均匀地分配到不同的 Pod 上，确保每个 Pod 都能接收到相同数量的请求，避免某些 Pod 过载而其他 Pod 空闲的状况。Kubernetes 常用的负载均衡方式包括 kube-proxy、Ingress 控制器以及使用外部负载均衡器等。

2. Kubernetes 的优势

(1) Kubernetes 系统不仅可以实现跨集群调度、水平扩展、监控、备份、灾难恢复，还可以解决大型互联网集群中多任务处理的问题。

(2) Kubernetes 遵循微服务架构理论，将整个系统划分为多个功能各异的组件。各组件结构清晰、部署简单，可以非常便捷地运行在系统环境中。

(3) 利用容器的扩容机制，系统将容器归类，形成“容器集”(Pod)，用于帮助用户调度工作负载，并为这些容器提供联网和存储服务。

9.1.2　Kubernetes 的体系架构

Kubernetes 集群主要由控制节点 Master 和多个工作节点 Node 组成，两种节点分别运行着不同的组件。Master 负责集群的管理，协调集群中的所有行为或活动，如应用的运行、修改、更新等；Node 负责接收 Master 的工作指令并执行相应的任务。集群状态存储系统 Etcd 为系统提供数据存储服务，Kubernetes 架构图如图 9-1 所示。

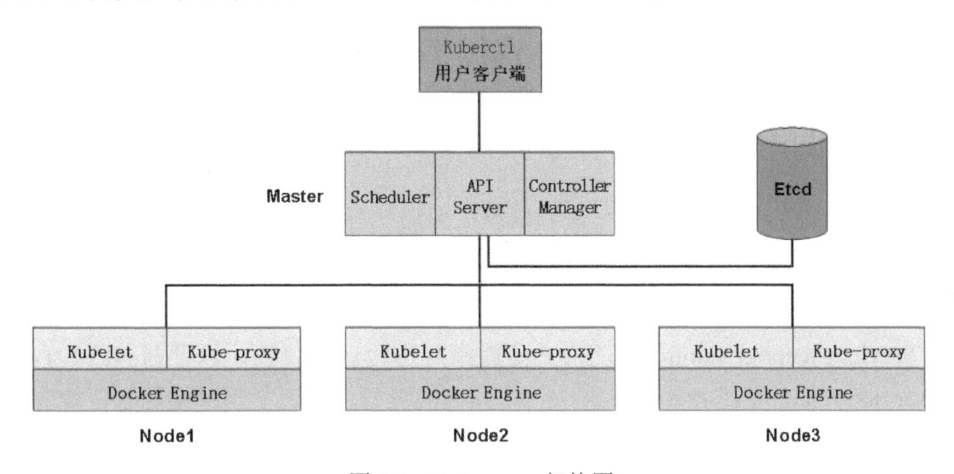

图 9-1　Kubernetes 架构图

1. Master 节点及其组件

控制节点 Maser 是整个集群的核心，主要负责组件或者服务进程的管理和控制，如追踪其他服务器的健康状态、保持各组件之间的通信、为用户或服务提供 API 接口等。Master 的所有组件通常会在一台主机上部署和启动，但是不在此主机上运行用户容器。

在控制节点 Master 上部署以下三种组件：

(1) API Server：整个系统的核心组件，是 Kubernetes 集群的入口，负责处理来自客户端的请求，包括对资源的增、删、改、查以及监控等操作。API Server 还通过 REST API 接口提供认证授权、数据校验以及集群状态变更等功能，是其他模块之间数据交互和通信的中枢。另外，API Server 是整个 Kubernetes 集群操作 Etcd 存储系统的唯一入口，确保集群数据的一致性和安全性。

(2) Scheduler：监视新创建且未分配工作节点的 Pod，根据不同的需求将其分配到工作节点中运行，同时允许用户根据实际需求定制调度策略，实现工作负载管理和资源调度，确保工作负载的合理分布和资源的有效利用。

(3) Controller Manager：集群的管理控制中心，负责管理集群中各种控制器(Controllers)，这些控制器是实现集群状态期望值的核心组件，负责监控和维护集群内各种资源对象的状态，包括Pods、Nodes、Deployments、Services、Endpoints等。

Kubernetes主要的控制器及其功能如下：

(1) Deployment Controller：管理一组Pod，确保指定数量的Pod在集群中运行。当有节点出现故障时，Deployment Controller会自动在其他节点上创建新的Pod以保持指定的Pod数量。另外，它还支持滚动更新和回滚操作，允许在不中断服务的情况下更新Pod的镜像或配置。

(2) Node Controller：管理和监控集群中的各个Node节点，如集群的Node信息同步、单个Node的生命周期管理、负载发现等。通过与API Server交互实时更新Node的状态信息，确保集群的高可用性和资源的优化利用。

(3) Namespace Controller：监控和管理Namespace资源对象，确保每个Namespace符合预期的状态。通过与API Server交互实时监控Namespace的状态，必要时进行相应的管理操作，确保整个集群的正常运行和资源的合理分配。

(4) Service Controller：负责与外部云平台进行交互，管理Service资源，如监听Service和对应Pod副本的变化、实现Service的负载均衡等，作为集群与外部云平台之间的接口确保Service资源的正确配置。

(5) Endpoints Controller：负责生成和维护所有的Endpoints对象，如监听Service和Pod的变化、当Service被删除时删除同名的Endpoints对象、实现负载均衡和故障恢复、与外部云平台进行交互等。

(6) Service Account Controller：为Pod提供身份验证和授权，以便与API Server进行交互。Service Account使Pod在集群中具有独立的身份，从而实现精细的权限控制和安全策略。

(7) Persistent Volume Controller：负责管理和控制持久化存储的分配和使用，如存储资源的分配、访问控制和安全保障、回收策略管理、动态扩展、与PVC绑定等。

(8) DaemonSet Controller：管理和控制DaemonSet Pod的创建、更新和删除，确保每个Node上至少运行一个DaemonSet Pod，以便实现相应服务。

(9) Job Controller：管理一次性任务或批处理作业，如文件压缩、数据处理等，一旦任务完成，Pod便结束运行。

(10) Pod Autoscaler Controller：实现Pod的自动伸缩，定时获取监控数据进行策略匹配，当满足条件时执行Pod的伸缩动作。

(11) Cloud Controller：嵌入了云平台的控制逻辑，使集群可以与云提供商进行交互。

2. Node节点及其组件

Node节点是集群中的工作节点，主要负责接收Master的工作指令并执行相应的任务，既可以是物理机也可以是虚拟机。当某个Node节点出现故障时，Master节点会将负载切换到其他工作节点上。

在 Node 节点上部署了以下三种组件：

(1) Kubelet：主要负责管控容器，首先从 API Server 接收 Pod 的创建请求，然后进行相关的启动和停止容器等操作。另外，Kubelet 会监控容器的运行状态，并将信息通知给 API Server，确保容器按照期望状态运行。

(2) Kube-proxy：从 API Server 获取 Service 信息并创建相关的代理服务，维护节点上的一些网络规则，实现集群内的客户端 Pod 访问 Service，或者是集群外的主机通过 NodePort 等方式访问 Service。

(3) Docker Engine：Docker 的守护进程，负责容器的创建和管理工作。

3. 集群状态存储系统 Etcd

Kubernetes 集群中所有的状态信息都存储于 Etcd 数据库中，Etcd 是一个独立的服务组件，以分布式键值存储在集群中，具有实现集群发现、共享配置等功能，同时还可以提供监听机制。当键值发生改变时，Etcd 通知 API Server 并通过 Watch API 向客户端输出。

Kubernetes 集群还支持 DNS、Web UI 等插件，用于提供更完备的集群功能。Kubernetes 是一个平台，不提供应用程序级服务，也不提供或授权配置语言，但用户可以通过任意形式的声明性规范来实现所需的功能。

9.1.3　Kubernetes 的相关概念

Kubernetes 作为容器化管理平台，其工作机制中涉及到很多核心概念，它们对于理解系统关键资源对象以及这些资源对象在系统中扮演的角色等非常重要，下面就来介绍与 Kuberntes 集群相关的概念和术语。

1. Pod

Pod 是 Kubernetes 集群进行管理的最小单位，一个 Pod 可以包含一个或多个相关容器，同一个 Pod 内的容器共享网络命名空间和存储资源。

2. Label

资源标识符，用来区分不同对象的属性。Label 的键值对 (Key:Value) 可以在对象创建前后进行添加和修改。用户可以通过给指定资源对象绑定一个或多个 Label 来灵活地管理资源，如资源的分配、调度、配置、部署等。

Selector 选择器通过匹配 Label 定义资源之间的关系，给某个资源对象定义一个 Label，便可以通过 Label Selector 查询和筛选拥有某些 Label 的资源对象。

3. Pause 容器

Pause 容器为其他容器提供生命周期的隔离和协调，同时还为 Pod 中的容器提供稳定的网络环境，确保容器网络连接的可靠性。

4. Replication Controller

Pod 的副本控制器简称 RC，是实现弹性伸缩、动态扩容和滚动升级的核心，它保证了集群中存在指定数量的 Pod 副本，如停止多余的 Pod 副本，或启动不足的 Pod 副本。

Deployment 是一个更高层次的 API 对象，用于管理 ReplicaSet 和 Pod，并提供声明式更新等功能。

5. StatefulSet

在 Kubernetes 集群中，StatefulSet 用于管理系统中有状态应用的集群，如 MysQL、MongoDB、ZooKeeper 集群等。这些集群中每个节点都有固定的 ID 号，集群中的成员通过 ID 号相互通信。为了能够在其他节点上恢复某个失败的节点，这种集群中的 Pod 需挂载到共享存储的磁盘上，在删除或启用 Pod 后，Pod 的名称和 IP 地址会发生改变。

StatefulSet 可以使 Pod 副本的名称和 IP 地址在整个生命周期中保持不变，从而使 Pod 副本按照固定的顺序启动、更新或者删除。StatefulSet 有唯一的网络标识符 (IP 地址)，适用于需要持久存储、有序部署、扩展、删除和滚动更新的应用程序。

6. Service

具备不同业务功能而又彼此独立的多个 Service 微服务单元构成了网站，服务之间通过 TCP/UDP 进行通信，从而形成了强大而又灵活的弹性网络。Service 提供一个或多个 Pod 实例的稳定访问地址，Frontend Pod 通过 Service 提供的入口访问一组 Pod 集群。当 Kubernetes 集群中存在 DNS 附件时，Service 服务会自动创建一个 DNS 名称用于服务发现，将外部流量引入集群内部，并将到达 Service 的请求分发到后端 Pod 对象上。

7. Job

计划任务 Job 是 Kubernetes 集群用来控制批处理型任务的 API 对象。批处理型任务与长期服务型业务的主要区别是批处理型任务的运行有头有尾，而长期服务型任务在用户不停止的情况下将永远运行。Job 管理的 Pod 成功完成任务就自动退出，任务成功完成的标志根据不同的 spec.completions 策略而不同，单 Pod 型任务有一个 Pod 成功就标志完成，定数成功型任务要保证有 N 个任务全部成功才标志完成，工作队列型任务根据应用确认的全局成功来标志完成。

8. Namespace

集群中存在许多资源对象，这些资源对象可以是不同的项目或用户等。命名空间 Namespace 将这些资源对象从逻辑上进行隔离并设定控制策略，以便不同的分组在共享整个集群资源时可以被分别管理。

9. Volume

存储卷 Volume 是集群中的共享存储资源，为应用服务提供存储空间。Volume 可以被 Pod 中的多个容器使用和挂载，也可以用于容器之间的数据共享。

10. Endpoint

Endpoint 主要用于标识服务进程的访问点，一般由容器端口号和 Pod 的 IP 地址构成。

9.1.4　Kubernetes 集群的管理

1. Kubernetes 中的资源类型

在 Kubernetes 中，所有的操作对象都被当作资源来进行管理，通过 API Server 组件提供的接口实现对资源的增、删、改、查等操作。资源配置清单是用来创建和管理资源的配置文件。

Kubernetes 中的资源非常丰富，常见的资源类型及其主要功能见表 9-1 所示。

表 9-1　常见的资源类型及其主要功能

资源类型	主 要 功 能	资 源 名 称
工作负载型	用于承载具体的工作负载，如应用程序的实例化运行	Pod、ReplicaSet、Deployment、StatefulSet、DaemonSet、Job、CronJob
服务发现及均衡型	负责服务发现和负载均衡，使得外部可以访问到内部的 Pod	Service，Ingress
配置与存储型	提供数据存储和配置管理	Volume、CSI、ConfigMap、Secret、DownwardAPI
集群级型	提供集群级别的管理和安全控制	Namespace、Node、Role、ClusterRole、RoleBinding、ClusterRoleBinding
元数据型	提供元数据相关的功能，如资源的限制和模板定义	HPA、PodTemplate、LimitRange

Pod 是 Kubernetes 中最小的部署和管理单位。一个 Pod 可以包含一个或多个容器，这些容器共享相同的网络和存储，并被部署到同一台物理或虚拟机器上。

Kubernetes 中的大多数资源都与 Pod 有关，如 Deployment 用于定义 Pod 的副本数量和升级策略，以确保应用程序在 Kubernetes 集群中持续运行，Deployment 还支持滚动升级和回退功能；Service 用于将一组 Pod 封装为一个统一的网络服务，并提供负载均衡功能，使其可以在 Kubernetes 集群内部或外部访问；ConfigMap 用于存储应用程序的配置数据，如环境变量、配置文件等，并将其注入到 Pod 中；Volume(存储卷) 被定义在 Pod 上，可以被 Pod 内多个容器挂载使用，Volume 与 Pod 的生命周期相同，当 Pod 内的容器终止或者重启时，Volume 中的数据不会丢失。

2. Kubernetes 中的配置文件

Kubernetes 中通常使用 YAML 格式来定义配置文件，配置文件一般包含以下四部分：

(1) apiVersion：指定 Kubernetes API 的版本。

(2) kind：定义该文件所描述的资源类型，如 Pod、Deployment、Service 等。

(3) metadata：指定资源的元数据信息 (属性)，如名称、命名空间，标签等。

(4) pec：定义资源的规格，包括容器镜像、端口映射、副本数、服务类型等。

一个 YAML 配置文件内可以同时定义多个资源。YAML 文件的语法规则在项目五 Docker Compose 的使用中已经介绍了，在此不再赘述。

3. Kubectl 命令行工具

Kubectl 是管理 Kubernetes 集群的客户端工具，用户可以通过 Kubectl 工具运行相应的命令实现对 Kubernetes 资源对象的创建、管理和调试等操作。kubectl 命令行工具使用灵活、功能强大，受到运维和开发人员的青睐。

Kubectl 命令的格式为：

```
kubectl [command] [type] [name] [flags]
```

(1) command：子命令，用于对 Kubernetes 集群中的一个或多个资源对象进行操作，如 create、get、apply、describe、delete 等。

(2) type：指定资源对象类型，区分大小写，可以指定单数、复数或缩写形式。例如，运行以下三条命令会输出相同的结果：

```
kubectl get pod app
kubectl get pods app
kubectl get po app
```

(3) name：指定资源对象的名称，区分大小写，若省略名称，则显示所有资源的详细信息。

(4) flags：指定可选的参数，例如，可使用 -s 或 --server 参数指定 Kubernetes API 服务器的地址和端口。

注意：从命令行指定的参数会覆盖默认值和相应的环境变量。

4. Kubectl 命令中常用的子命令

run：基于特定镜像在 Kubernetes 集群中创建 Pod、Deployment 或 Job。

get：显示一个或者多个资源对象的信息。

logs：显示容器的日志。

proxy：将本机的某个端口映射到 API Server。

label：设置或者更新资源对象的 label。

apply：从 stdin 或者配置文件对资源对象更新配置。

create：从 stdin 或者配置文件创建资源对象。

delete：删除资源对象。

describe：描述一个或者多个资源对象的详细信息。

diff：查看配置文件与当前系统中正在运行的资源对象之间的差异。

edit：编辑资源对象的属性，在线更新。

exec：执行一个容器内的命令。

cluster-info：显示集群中 Master 和内置服务的信息。

config：修改 kubeconfig 文件。

set：设置资源对象的某个特定信息。

top：查看 Node 或 Pod 的资源使用情况。

version：显示系统版本信息。

scale：扩容或缩容 Deployment、ReplicaSet、RC 或者 Job 中 Pod 的数量。

plugin：在 Kubectl 命令行使用自定义插件。

9.2 任务实施

9.2.1 配置 Kubernetes 集群各节点的系统环境

配置 Kubernetes 集群各节点的系统环境

1. 任务目标

掌握配置 Kubernetes 集群各节点系统环境的方法。

2. 任务内容

(1) 克隆三台主机，并设置其 IP 地址。

(2) 各主机连接远程管理工具 MobaXterm。

(3) 设置各主机的主机名称。

(4) 配置各主机的主机名解析。

(5) 关闭系统 Swap。

3. 完成任务所需的设备和软件

(1) 一台安装 Windows 10 操作系统的计算机。

(2) VMware Workstation、Docker。

(3) 远程管理工具 MobaXterm。

4. 任务实施步骤

第一步，利用 CentOS7-m 虚拟机克隆三台服务器，虚拟机名称分别为 master、node1 和 node2，Kubernetes 集群的基本设置信息见表 9-2 所示。

表 9-2　Kubernetes 集群的基本设置信息

节点名称	IP 地址	操作系统	主要软件	CPU 配置	内存设置
master	192.168.1.20	CentOS 7.6	Docker CE	2 核	2 GB
node1	192.168.1.21	CentOS 7.6	Docker CE	2 核	2 GB
node2	192.168.1.22	CentOS 7.6	Docker CE	2 核	2 GB

以下步骤需在 master、node1 和 node2 三台节点上同时运行，在此仅给出 master 节点的运行情况，其他节点可参照进行。

第二步，开启 master 节点，修改其 IP 地址为 192.168.1.20 并重启网络，操作命令如下：

```
[root@docker ~]# vi /etc/sysconfig/network-scripts/ifcfg-ens33
TYPE="Ethernet"
BOOTPROTO="static"
IPADDR=192.168.1.20
NETMASK=255.255.255.0
```

```
GATEWAY=192.168.1.2
NAME="ens33"
DEVICE="ens33"
ONBOOT="yes"
DNS1=114.114.114.114
DNS2=8.8.8.8
[root@docker ~]# systemctl restart network
```

第三步，通过远程管理工具 MobaXterm 连接 master 节点，如图 9-2 所示。

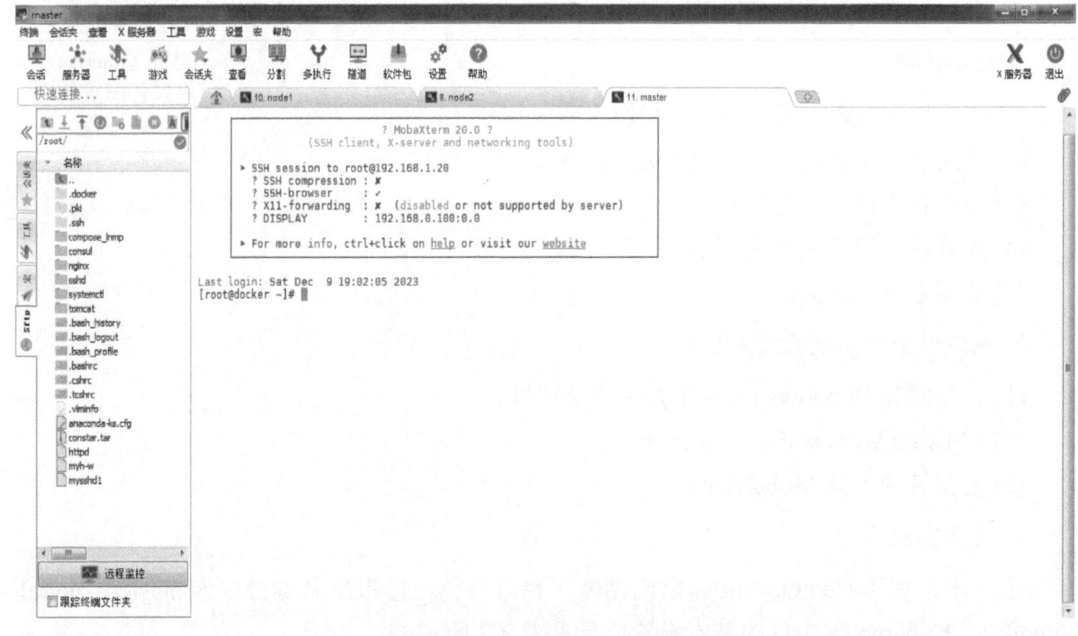

图 9-2 远程管理工具 MobaXterm 连接 Master 节点

第四步，在 master 节点上，修改其主机名为 master，操作命令如下：

```
[root@docker ~]# hostnamectl set-hostname master
[root@docker ~]# su
[root@master ~]# hostname
```

命令运行结果如图 9-3 所示。

```
[root@docker ~]# hostnamectl set-hostname master
[root@docker ~]# su
[root@master ~]# hostname
master
[root@master ~]#
```

图 9-3 修改 Master 节点的主机名

第五步，在 master 节点上，配置各节点间的主机名解析，操作命令如下：

```
[root@master ~]# vim /etc/hosts
127.0.0.1   localhost localhost.localdomain localhost4 localhost4.localdomain4
```

::1　　localhost localhost.localdomain localhost6 localhost6.localdomain6

192.168.1.20 master

192.168.1.21 node1

192.168.1.22 node2

第六步，在 master 节点上，关闭 Swap(交换分区) 系统，操作命令如下：

[root@master ~]# free –m

[root@master ~]# swapoff –a

[root@master ~]# free –m

命令运行结果如图 9-4 所示。

```
[root@master ~]# free -m
              total        used        free      shared  buff/cache   available
Mem:           1819         161        1317           9         340        1455
Swap:          2047           0        2047
[root@master ~]# swapoff -a
[root@master ~]# free -m
              total        used        free      shared  buff/cache   available
Mem:           1819         160        1320           9         338        1458
Swap:             0           0           0
[root@master ~]#
```

图 9-4　在 master 节点上关闭 Swap 系统

第七步，永久关闭 Swap 系统，注释掉 Swap 自动挂载配置，否则默认配置下的 Kubelet 将无法正常启动，操作命令如下：

[root@master ~]# vim /etc/fstab

操作结果如图 9-5 所示。

```
#
# /etc/fstab
# Created by anaconda on Fri Apr  1 10:39:53 2022
#
# Accessible filesystems, by reference, are maintained under '/dev/disk'
# See man pages fstab(5), findfs(8), mount(8) and/or blkid(8) for more info
#
/dev/mapper/centos-root /                       xfs     defaults        0 0
UUID=730d6771-0e80-4020-a1a7-db6b9d39aff8 /boot                   xfs     defaults        0 0
#/dev/mapper/centos-swap swap                    swap    defaults        0 0
```

图 9-5　注释掉 Swap 自动挂载配置 (永久关闭 Swap)

第八步，根据需要为 master 节点创建快照，保存此时的状态。

9.2.2　部署 Kubernetes 集群

部署 Kubernetes 集群

1. 任务目标

掌握部署 Kubernetes 集群的方法。

2. 任务内容

(1) 配置 Kubeadm 和 Kubelet 的 Repo 源并安装。

(2) 配置网络转发参数并使其生效。

(3) 加载 IPVS 相关内核模块。

(4) 配置 Kubelet 的 Cgroups，启动 Kubelet 服务。

(5) 初始化 Master 节点。

(6) 安装网络插件 flannel 并启动 Kubelet。

(7) 将 Node1 和 Node2 加入集群。

3. 完成任务所需的设备和软件

(1) 一台安装 Windows 10 操作系统的计算机。

(2) VMware Workstation、Docker。

(3) 远程管理工具 MobaXterm。

4. 任务实施步骤

以下第一步至第六步需在 master、node1 和 node2 三台节点上同时运行，在此仅给出 master 节点的运行步骤，其他节点可参照进行。

第一步，配置 Kubeadm 和 Kubelet 的 Repo 源，下载安装信息并缓存到本地，操作命令如下：

```
[root@master ~]# vim /etc/yum.repos.d/Kubernetes.repo
[Kubernetes]
name=Kubernetes
baseurl=https://mirrors.aliyun.com/kubernetes/yum/repos/kubernetes-el7-x86_64
enabled=1
gpgcheck=0
repo_gpgcheck=0
gpgkey=https://mirrors.aliyun.com/kubernetes/yum/doc/yum-key.gpg https://mirrors.aliyun.com/kubernetes/yum/doc/rpm-package-key.gpg
[root@master ~]# yum makecache fast
```

命令运行结果如图 9-6 所示。

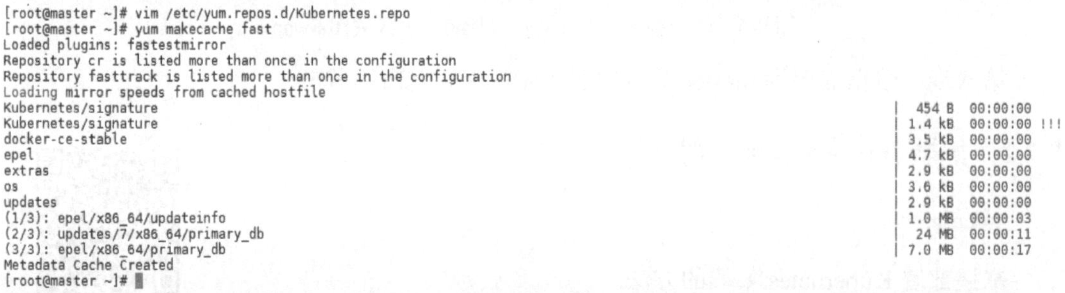

```
[root@master ~]# vim /etc/yum.repos.d/Kubernetes.repo
[root@master ~]# yum makecache fast
Loaded plugins: fastestmirror
Repository cr is listed more than once in the configuration
Repository fasttrack is listed more than once in the configuration
Loading mirror speeds from cached hostfile
Kubernetes/signature                                   |  454 B  00:00:00
Kubernetes/signature                                   |  1.4 kB  00:00:00 !!!
docker-ce-stable                                       |  3.5 kB  00:00:00
epel                                                   |  4.7 kB  00:00:00
extras                                                 |  2.9 kB  00:00:00
os                                                     |  3.6 kB  00:00:00
updates                                                |  2.9 kB  00:00:00
(1/3): epel/x86_64/updateinfo                          |  1.0 MB  00:00:03
(2/3): updates/7/x86_64/primary_db                     |   24 MB  00:00:11
(3/3): epel/x86_64/primary_db                          |  7.0 MB  00:00:17
Metadata Cache Created
[root@master ~]#
```

图 9-6　配置 Kubeadm 和 Kubelet 的 Repo 源，下载安装信息并缓存到本地

第二步，安装 Kubeadm 和 Kubelet 工具，操作命令如下：

```
[root@master ~]# yum install -y kubelet-1.19.0 kubeadm-1.19.0 kubectl-1.19.0
```

命令运行结果如图 9-7 所示。

```
[root@master ~]# yum install -y kubelet-1.19.0 kubeadm-1.19.0 kubectl-1.19.0
Loaded plugins: fastestmirror
Repository cr is listed more than once in the configuration
Repository fasttrack is listed more than once in the configuration
Loading mirror speeds from cached hostfile
Resolving Dependencies
--> Running transaction check
---> Package kubeadm.x86_64 0:1.19.0-0 will be installed
--> Processing Dependency: kubernetes-cni >= 0.8.6 for package: kubeadm-1.19.0-0.x86_64
---> Package kubectl.x86_64 0:1.19.0-0 will be installed
---> Package kubelet.x86_64 0:1.19.0-0 will be installed
--> Running transaction check
---> Package kubernetes-cni.x86_64 0:1.2.0-0 will be installed
--> Finished Dependency Resolution
        ……                                    ……
Transaction test succeeded
Running transaction
  Installing : kubelet-1.19.0-0.x86_64                                   1/4
  Installing : kubernetes-cni-1.2.0-0.x86_64                             2/4
  Installing : kubectl-1.19.0-0.x86_64                                   3/4
  Installing : kubeadm-1.19.0-0.x86_64                                   4/4
  Verifying  : kubeadm-1.19.0-0.x86_64                                   1/4
  Verifying  : kubernetes-cni-1.2.0-0.x86_64                             2/4
  Verifying  : kubectl-1.19.0-0.x86_64                                   3/4
  Verifying  : kubelet-1.19.0-0.x86_64                                   4/4

Installed:
  kubeadm.x86_64 0:1.19.0-0          kubectl.x86_64 0:1.19.0-0          kubelet.x86_64 0:1.19.0-0

Dependency Installed:
  kubernetes-cni.x86_64 0:1.2.0-0

Complete!
```

图 9-7　安装 Kubeadm 和 Kubelet 工具

第三步，配置网络转发参数并使其生效，确保集群能够正常通信，操作命令如下：

```
[root@master ~]# vim /etc/sysctl.d/k8s.conf
net.bridge.bridge-nf-call-ip6tables = 1
net.bridge.bridge-nf-call-iptables = 1
vm.swappiness = 0
[root@master ~]# sysctl –system
```

命令运行结果如图 9-8 所示。

```
[root@master ~]#  vim /etc/sysctl.d/k8s.conf
[root@master ~]# sysctl --system
* Applying /usr/lib/sysctl.d/00-system.conf ...
net.bridge.bridge-nf-call-ip6tables = 0
net.bridge.bridge-nf-call-iptables = 0
net.bridge.bridge-nf-call-arptables = 0
* Applying /usr/lib/sysctl.d/10-default-yama-scope.conf ...
kernel.yama.ptrace_scope = 0
* Applying /usr/lib/sysctl.d/50-default.conf ...
kernel.sysrq = 16
kernel.core_uses_pid = 1
net.ipv4.conf.default.rp_filter = 1
net.ipv4.conf.all.rp_filter = 1
net.ipv4.conf.default.accept_source_route = 0
net.ipv4.conf.all.accept_source_route = 0
net.ipv4.conf.default.promote_secondaries = 1
net.ipv4.conf.all.promote_secondaries = 1
fs.protected_hardlinks = 1
fs.protected_symlinks = 1
* Applying /etc/sysctl.d/99-sysctl.conf ...
* Applying /etc/sysctl.d/k8s.conf ...
net.bridge.bridge-nf-call-ip6tables = 1
net.bridge.bridge-nf-call-iptables = 1
vm.swappiness = 0
* Applying /etc/sysctl.conf ...
```

图 9-8　配置网络转发参数并使其生效

第四步，加载 IPVS 相关内核模块，并查看是否加载成功，操作命令如下：

```
[root@master ~]# modprobe ip_vs
[root@master ~]# modprobe ip_vs_rr
[root@master ~]# modprobe ip_vs_wrr
[root@master ~]# modprobe ip_vs_sh
[root@master ~]# modprobe nf_conntrack_ipv4
```

```
[root@master ~]# lsmod | grep ip_vs
```

命令运行结果如图 9-9 所示。

```
[root@master ~]# modprobe ip_vs
[root@master ~]# modprobe ip_vs_rr
[root@master ~]# modprobe ip_vs_wrr
[root@master ~]# modprobe ip_vs_sh
[root@master ~]# modprobe nf_conntrack_ipv4
[root@master ~]# lsmod | grep ip_vs
ip_vs_sh               12688  0
ip_vs_wrr              12697  0
ip_vs_rr               12600  0
ip_vs                 145497  6 ip_vs_rr,ip_vs_sh,ip_vs_wrr
nf_conntrack          133095  7 ip_vs,nf_nat,nf_nat_ipv4,xt_conntrack,nf_nat_masquerade_ipv4,nf_conntrack_netlink,nf_conntrack_ipv4
libcrc32c              12644  4 xfs,ip_vs,nf_nat,nf_conntrack
[root@master ~]#
```

图 9-9　加载 IPVS 相关内核模块并查看是否加载成功

第五步，获取 Docker 的 Cgroups，配置 Kubelet 的 Cgroups，操作命令如下：

```
[root@master ~]# DOCKER_CGROUPS=$(docker info | grep 'Cgroup' | cut -d' ' -f4)
[root@master ~]# echo $DOCKER_CGROUPS
[root@master ~]# cat >/etc/sysconfig/kubelet<<EOF
KUBELET_EXTRA_ARGS="--cgroup-driver=$DOCKER_CGROUPS --pod-infra-container-image=registry.cn-hangzhou.aliyuncs.com/google_containers/pause-amd64:3.1"
EOF
```

如果获取 Docker 的 Cgroups 时出现 WARNING: IPv4 forwarding is disabled，则需重启网络和 docker，命令运行结果如图 9-10 所示。

```
[root@master ~]# DOCKER_CGROUPS=$(docker info | grep 'Cgroup' | cut -d' ' -f4)
WARNING: IPv4 forwarding is disabled
[root@master ~]# systemctl restart network
[root@master ~]# systemctl restart docker
[root@master ~]# DOCKER_CGROUPS=$(docker info | grep 'Cgroup' | cut -d' ' -f4)
[root@master ~]# echo $DOCKER_CGROUPS
cgroupfs 1
[root@master ~]# cat >/etc/sysconfig/kubelet<<EOF
KUBELET_EXTRA_ARGS="--cgroup-driver=$DOCKER_CGROUPS --pod-infra-container-image=registry.cn-hangzhou.aliyuncs.com/google_containers/pause-
amd64:3.1"
EOF
[root@master ~]#
```

图 9-10　配置 Kubelet 的 Cgroups

第六步，启动 Kubelet 服务，并查看其状态，操作命令如下：

```
[root@master ~]# systemctl daemon-reload
[root@master ~]# systemctl restart kubelet
[root@master ~]# systemctl enable kubelet
[root@master ~]# systemctl status kubelet
```

命令运行结果如图 9-11 所示。

```
[root@master ~]# systemctl daemon-reload
[root@master ~]# systemctl restart kubelet
[root@master ~]# systemctl enable kubelet
Created symlink from /etc/systemd/system/multi-user.target.wants/kubelet.service to /usr/lib/systemd/system/kubelet.service.
[root@master ~]# systemctl status kubelet
● kubelet.service - kubelet: The Kubernetes Node Agent
   Loaded: loaded (/usr/lib/systemd/system/kubelet.service; enabled; vendor preset: disabled)
  Drop-In: /usr/lib/systemd/system/kubelet.service.d
           └─10-kubeadm.conf
   Active: activating (auto-restart) (Result: exit-code) since Thu 2023-12-14 16:52:00 CST; 8s ago
     Docs: https://kubernetes.io/docs/
  Process: 19428 ExecStart=/usr/bin/kubelet $KUBELET_KUBECONFIG_ARGS $KUBELET_CONFIG_ARGS $KUBELET_KUBEADM_ARGS $KUBELET_EXTRA_ARGS (code=
exited, status=1/FAILURE)
 Main PID: 19428 (code=exited, status=1/FAILURE)

Dec 14 16:52:00 node1 systemd[1]: kubelet.service: main process exited, code=exited, status=1/FAILURE
Dec 14 16:52:00 node1 systemd[1]: Unit kubelet.service entered failed state.
Dec 14 16:52:00 node1 systemd[1]: kubelet.service failed.
[root@master ~]#
```

图 9-11　启动 Kubelet 服务，并查看其状态

第七步，初始化 master 节点，操作命令如下：

[root@master ~]# kubeadm init --kubernetes-version=v1.19.0 --service-cidr=10.96.0.0/12 --pod-network-cidr=10.244.0.0/16 --apiserver-advertise-address=192.168.1.20 --ignore-preflight-errors=Swap --image-repository=registry.aliyuncs.com/google_containers

命令运行后，会显示出以下内容：

W1214 21:26:13.945273 28698 configset.go:348] WARNING: kubeadm cannot validate component configs for API groups [kubelet.config.k8s.io kubeproxy.config.k8s.io]

[init] Using Kubernetes version: v1.19.0

[preflight] Running pre-flight checks

[WARNING IsDockerSystemdCheck]: detected "cgroupfs" as the Docker cgroup driver. The recommended driver is "systemd". Please follow the guide at https://kubernetes.io/docs/setup/cri/

[WARNING SystemVerification]: this Docker version is not on the list of validated versions: 20.10.14. Latest validated version: 19.03

[preflight] Pulling images required for setting up a Kubernetes cluster

[preflight] This might take a minute or two, depending on the speed of your internet connection

[preflight] You can also perform this action in beforehand using 'kubeadm config images pull'

[certs] Using certificateDir folder "/etc/kubernetes/pki"

[certs] Generating "ca" certificate and key

[certs] Generating "apiserver" certificate and key

[certs] apiserver serving cert is signed for DNS names [kubernetes kubernetes.default kubernetes.default.svc kubernetes.default.svc.cluster.local master] and IPs [10.96.0.1 192.168.1.20]

[certs] Generating "apiserver-kubelet-client" certificate and key

[certs] Generating "front-proxy-ca" certificate and key

[certs] Generating "front-proxy-client" certificate and key

[certs] Generating "etcd/ca" certificate and key

[certs] Generating "etcd/server" certificate and key

[certs] etcd/server serving cert is signed for DNS names [localhost master] and IPs [192.168.1.20 127.0.0.1 ::1]

[certs] Generating "etcd/peer" certificate and key

[certs] etcd/peer serving cert is signed for DNS names [localhost master] and IPs [192.168.1.20 127.0.0.1 ::1]

[certs] Generating "etcd/healthcheck-client" certificate and key

[certs] Generating "apiserver-etcd-client" certificate and key

[certs] Generating "sa" key and public key

[kubeconfig] Using kubeconfig folder "/etc/kubernetes"

[kubeconfig] Writing "admin.conf" kubeconfig file

[kubeconfig] Writing "kubelet.conf" kubeconfig file

[kubeconfig] Writing "controller-manager.conf" kubeconfig file

[kubeconfig] Writing "scheduler.conf" kubeconfig file

[kubelet-start] Writing kubelet environment file with flags to file "/var/lib/kubelet/kubeadm-flags.env"

[kubelet-start] Writing kubelet configuration to file "/var/lib/kubelet/config.yaml"

```
[kubelet-start] Starting the kubelet
[control-plane] Using manifest folder "/etc/kubernetes/manifests"
[control-plane] Creating static Pod manifest for "kube-apiserver"
[control-plane] Creating static Pod manifest for "kube-controller-manager"
[control-plane] Creating static Pod manifest for "kube-scheduler"
[etcd] Creating static Pod manifest for local etcd in "/etc/kubernetes/manifests"
[wait-control-plane] Waiting for the kubelet to boot up the control plane as static Pods from directory "/etc/kubernetes/manifests". This can take up to 4m0s
[apiclient] All control plane components are healthy after 36.180436 seconds
[upload-config] Storing the configuration used in ConfigMap "kubeadm-config" in the "kube-system" Namespace
[kubelet] Creating a ConfigMap "kubelet-config-1.19" in namespace kube-system with the configuration for the kubelets in the cluster
[upload-certs] Skipping phase. Please see --upload-certs
[mark-control-plane] Marking the node master as control-plane by adding the label "node-role.kubernetes.io/master=''"
[mark-control-plane] Marking the node master as control-plane by adding the taints [node-role.kubernetes.io/master:NoSchedule]
[bootstrap-token] Using token: on2k3m.xpk70r1pmjrcemw1
[bootstrap-token] Configuring bootstrap tokens, cluster-info ConfigMap, RBAC Roles
[bootstrap-token] configured RBAC rules to allow Node Bootstrap tokens to get nodes
[bootstrap-token] configured RBAC rules to allow Node Bootstrap tokens to post CSRs in order for nodes to get long term certificate credentials
[bootstrap-token] configured RBAC rules to allow the csrapprover controller automatically approve CSRs from a Node Bootstrap Token
[bootstrap-token] configured RBAC rules to allow certificate rotation for all node client certificates in the cluster
[bootstrap-token] Creating the "cluster-info" ConfigMap in the "kube-public" namespace
[kubelet-finalize] Updating "/etc/kubernetes/kubelet.conf" to point to a rotatable kubelet client certificate and key
[addons] Applied essential addon: CoreDNS
[addons] Applied essential addon: kube-proxy

Your Kubernetes control-plane has initialized successfully!

To start using your cluster, you need to run the following as a regular user:

mkdir -p $HOME/.kube
sudo cp -i /etc/kubernetes/admin.conf $HOME/.kube/config
```

```
sudo chown $(id -u):$(id -g) $HOME/.kube/config

You should now deploy a pod network to the cluster.
Run "kubectl apply -f [podnetwork].yaml" with one of the options listed at:
https://kubernetes.io/docs/concepts/cluster-administration/addons/

Then you can join any number of worker nodes by running the following on each as root:

kubeadm join 192.168.1.20:6443 --token on2k3m.xpk70r1pmjrcemw1 \
    --discovery-token-ca-cert-hash sha256:ec388416ec8c934948157963b39fcbd2d56bf143ea15b2bfd119b
e143194d683
```

注意：最后三行代码是配置 Node 节点加入集群的 token 指令，在 Node 节点输入此 token 指令即可加入集群，请将其复制保存。

Kubernetes 集群初始化完成了四项工作：

(1) [kubelet]：生成 Kubelet 的配置文件“/var/lib/kubelet/config.yaml”。

(2) [certificates]：生成相关的各种证书。

(3) [kubeconfig]：生成 kubeconfig 文件。

(4) [bootstraptoken]：生成 token。

第八步，在 master 节点配置 Kubectl，并查看节点状态，操作命令如下：

```
[root@master ~]# rm -rf $HOME/.kube
[root@master ~]# mkdir -p $HOME/.kube
[root@master ~]# cp -i /etc/kubernetes/admin.conf $HOME/.kube/config
[root@master ~]# chown $(id -u):$(id -g) $HOME/.kube/config
[root@master ~]# kubectl get nodes
```

命令运行结果如图 9-12 所示。

```
[root@master ~]# rm -rf $HOME/.kube
[root@master ~]# mkdir -p $HOME/.kube
[root@master ~]# ls -a $HOME/
.   anaconda-ks.cfg  .bash_logout   .bashrc        constar.tar  .cshrc   httpd  myh-w    nginx  .ssh  systemctl  tomcat
..  .bash_history    .bash_profile  compose_lnmp   consul       .docker  .kube  mysshd1  .pki   sshd  .tcshrc    .viminfo
[root@master ~]# cp -i /etc/kubernetes/admin.conf $HOME/.kube/config
[root@master ~]# ls  $HOME/.kube/
config
[root@master ~]# vim $HOME/.kube/config
[root@master ~]# chown $(id -u):$(id -g) $HOME/.kube/config
[root@master ~]# kubectl get nodes
NAME     STATUS     ROLES    AGE   VERSION
master   NotReady   master   10h   v1.19.0
```

图 9-12　在 master 节点配置 Kubectl，并查看节点状态

注意：此时 master 节点的状态为“NotReady”，因为还没有安装网络插件。

第九步，在 master 节点上，安装网络插件 flannel，并启动 Kubectl，操作命令如下：

```
[root@master ~]# docker pull quay.io/coreos/flannel:v0.11.0-amd64
[root@master ~]# wget https://raw.githubusercontent.com/coreos/flannel/master/ Documentation/kube-
flannel.yml
[root@master ~]# kubectl apply -f kube-flannel.yml
```

命令运行结果如图9-13所示。

```
[root@master ~]# docker pull quay.io/coreos/flannel:v0.11.0-amd64
v0.11.0-amd64: Pulling from coreos/flannel
cd784148e348: Pull complete
04ac94e9255c: Pull complete
e10b013543eb: Pull complete
005e31e443b1: Pull complete
74f794f05817: Pull complete
Digest: sha256:7806805c93b20a168d0bbbd25c6a213f00ac58a511c47e8fa6409543528a204e
Status: Downloaded newer image for quay.io/coreos/flannel:v0.11.0-amd64
quay.io/coreos/flannel:v0.11.0-amd64
[root@master ~]# wget https://raw.githubusercontent.com/coreos/flannel/master/Documentation/kube-flannel.yml
--2023-12-15 11:09:48--  https://raw.githubusercontent.com/coreos/flannel/master/Documentation/kube-flannel.yml
Resolving raw.githubusercontent.com (raw.githubusercontent.com)... 185.199.108.133, 185.199.109.133, 185.199.110.133, ...
Connecting to raw.githubusercontent.com (raw.githubusercontent.com)|185.199.108.133|:443... connected.
HTTP request sent, awaiting response... 200 OK
Length: 4398 (4.3K) [text/plain]
Saving to: 'kube-flannel.yml'

100%[==========================================================================================>] 4,398       18.2KB/s   in 0.2s

2023-12-15 11:09:52 (18.2 KB/s) - 'kube-flannel.yml' saved [4398/4398]

[root@master ~]# ls
anaconda-ks.cfg  compose_lnmp  constar.tar  consul  httpd  kube-flannel.yml  myh-w  mysshd1  nginx  sshd  systemctl  tomcat
[root@master ~]# kubectl apply -f kube-flannel.yml
namespace/kube-flannel created
clusterrole.rbac.authorization.k8s.io/flannel configured
clusterrolebinding.rbac.authorization.k8s.io/flannel configured
serviceaccount/flannel created
configmap/kube-flannel-cfg created
daemonset.apps/kube-flannel-ds created
```

图9-13　在master节点上安装网络插件flannel并启动Kubectl

第十步，在node1节点和node2节点上，安装网络插件flannel，并启动Kubectl，操作命令如下：

注意：此处仅给出node1节点的操作，node2节点的操作类似。

```
[root@node1 ~]# docker pull quay.io/coreos/flannel:v0.11.0-amd64
[root@node1 ~]# wget https://raw.githubusercontent.com/coreos/flannel/master/ Documentation/kube-flannel.yml
[root@node1 ~]# kubectl apply -f kube-flannel.yml
```

命令运行结果如图9-14所示。

```
[root@node1 ~]# docker pull quay.io/coreos/flannel:v0.11.0-amd64
v0.11.0-amd64: Pulling from coreos/flannel
cd784148e348: Pull complete
04ac94e9255c: Pull complete
e10b013543eb: Pull complete
005e31e443b1: Pull complete
74f794f05817: Pull complete
Digest: sha256:7806805c93b20a168d0bbbd25c6a213f00ac58a511c47e8fa6409543528a204e
Status: Downloaded newer image for quay.io/coreos/flannel:v0.11.0-amd64
quay.io/coreos/flannel:v0.11.0-amd64
[root@node1 ~]# wget https://raw.githubusercontent.com/coreos/flannel/master/Documentation/kube-flannel.yml
--2023-12-15 16:49:43--  https://raw.githubusercontent.com/coreos/flannel/master/Documentation/kube-flannel.yml
Resolving raw.githubusercontent.com (raw.githubusercontent.com)... 185.199.111.133, 185.199.109.133, 185.199.110.133, ...
Connecting to raw.githubusercontent.com (raw.githubusercontent.com)|185.199.111.133|:443... connected.
HTTP request sent, awaiting response... 200 OK
Length: 4398 (4.3K) [text/plain]
Saving to: 'kube-flannel.yml'

100%[==========================================================================================>] 4,398       --.-K/s   in 0.002s

2023-12-15 16:49:45 (1.79 MB/s) - 'kube-flannel.yml' saved [4398/4398]

[root@node1 ~]# kubectl apply -f kube-flannel.yml
The connection to the server localhost:8080 was refused - did you specify the right host or port?
[root@node1 ~]#
```

图9-14　在node1节点上安装网络插件flannel并启动Kubectl

注意：以上启动Kubectl失败，原因是kubectl命令需要使用kubernetes-admin的身份来运行，在“kubeadm int”启动集群的步骤中已经生成“/etc/kubernetes/admin.conf”，只有将主节点中的“/etc/kubernetes/admin.conf”文件拷贝到node1节点和node2节点相同的目录下，并且配置环境变量，才能启动Kubectl成功。解决办法如下：

(1) 首先需在master节点上复制文件到node1和node2节点，操作命令如下：

```
[root@master ~]# scp /etc/kubernetes/admin.conf root@192.168.1.21:/etc/ kubernetes/admin.conf
```

[root@master ~]# scp /etc/kubernetes/admin.conf root@192.168.1.22:/etc/ kubernetes/admin.conf

命令运行结果如图 9-15 所示。

```
[root@master ~]# scp /etc/kubernetes/admin.conf root@192.168.1.21:/etc/kubernetes/admin.conf
The authenticity of host '192.168.1.21 (192.168.1.21)' can't be established.
ECDSA key fingerprint is SHA256:zmybdvjAZHC05RbamaOhXATLTW3uJzAmXPM1xKhIciA.
ECDSA key fingerprint is MD5:db:2c:50:fc:56:c5:71:45:08:3f:68:72:77:6a:b3:d2.
Are you sure you want to continue connecting (yes/no)? yes
Warning: Permanently added '192.168.1.21' (ECDSA) to the list of known hosts.
root@192.168.1.21's password:
admin.conf                                                    100% 5568   139.3KB/s   00:00
[root@master ~]# scp /etc/kubernetes/admin.conf root@192.168.1.22:/etc/kubernetes/admin.conf
The authenticity of host '192.168.1.22 (192.168.1.22)' can't be established.
ECDSA key fingerprint is SHA256:zmybdvjAZHC05RbamaOhXATLTW3uJzAmXPM1xKhIciA.
ECDSA key fingerprint is MD5:db:2c:50:fc:56:c5:71:45:08:3f:68:72:77:6a:b3:d2.
Are you sure you want to continue connecting (yes/no)? yes
Warning: Permanently added '192.168.1.22' (ECDSA) to the list of known hosts.
root@192.168.1.22's password:
admin.conf                                                    100% 5568   139.3KB/s   00:00
```

图 9-15　在 master 节点上复制文件到 node1 和 node2 节点

(2) 在 node1 和 node2 节点上配置环境变量，并启动 Kubectl，此处以 node1 为例，操作命令如下：

[root@node1 ~]# export KUBECONFIG=/etc/kubernetes/admin.conf

[root@node1 ~]# echo "export KUBECONFIG=/etc/kubernetes/admin.conf" >> ~/.bash_profile

[root@node1 ~]# kubectl apply -f kube-flannel.yml

命令运行结果如图 9-16 所示。

```
[root@node1 ~]# export KUBECONFIG=/etc/kubernetes/admin.conf
[root@node1 ~]# echo "export KUBECONFIG=/etc/kubernetes/admin.conf" >> ~/.bash_profile
[root@node1 ~]# kubectl apply -f kube-flannel.yml
namespace/kube-flannel unchanged
clusterrole.rbac.authorization.k8s.io/flannel unchanged
clusterrolebinding.rbac.authorization.k8s.io/flannel unchanged
serviceaccount/flannel unchanged
configmap/kube-flannel-cfg unchanged
daemonset.apps/kube-flannel-ds unchanged
```

图 9-16　在 node1 节点上配置环境变量并启动 Kubectl

第十一步，在 master 节点查看 Kubectl 信息及节点状态，操作命令如下：

[root@master ~]# kubectl get pods --namespace kube-system

[root@master ~]# kubectl get service

[root@master ~]# kubectl get svc --namespace kube-system

[root@master ~]# kubectl get nodes

命令运行结果如图 9-17 所示。

```
[root@master ~]# kubectl get pods --namespace kube-system
NAME                              READY   STATUS    RESTARTS   AGE
coredns-6d56c8448f-tbwfn          1/1     Running   0          13h
coredns-6d56c8448f-tmk74          1/1     Running   0          13h
etcd-master                       1/1     Running   0          13h
kube-apiserver-master             1/1     Running   0          13h
kube-controller-manager-master    1/1     Running   1          13h
kube-proxy-mkrbk                  1/1     Running   0          13h
kube-scheduler-master             1/1     Running   1          13h
[root@master ~]# kubectl get service
NAME         TYPE        CLUSTER-IP   EXTERNAL-IP   PORT(S)   AGE
kubernetes   ClusterIP   10.96.0.1    <none>        443/TCP   13h
[root@master ~]# kubectl get svc --namespace kube-system
NAME       TYPE        CLUSTER-IP    EXTERNAL-IP   PORT(S)                  AGE
kube-dns   ClusterIP   10.96.0.10    <none>        53/UDP,53/TCP,9153/TCP   13h
[root@master ~]# kubectl get nodes
NAME     STATUS   ROLES    AGE   VERSION
master   Ready    master   13h   v1.19.0
```

图 9-17　在 master 节点查看 Kubectl 信息及节点状态

第十二步，在 node1 和 node2 节点分别输入 token 指令，将他们加入集群，此处以 node1 节点为例，操作命令如下：

```
[root@node1 ~]# kubeadm join 192.168.1.20:6443 –token on2k3m.xpk70r1pmjrcemw1 --discovery-token-ca-cert-hash sha256:ec388416ec8c934948157963b39fcbd2d56bf143ea15 b2bfd119be143194d683
```

注意：将之前复制的 token 指令粘贴过来即可。

命令运行结果如图 9-18 所示。

```
[root@node1 ~]# kubeadm join 192.168.1.20:6443 --token on2k3m.xpk70r1pmjrcemw1 \
>     --discovery-token-ca-cert-hash sha256:ec388416ec8c934948157963b39fcbd2d56bf143ea15b2bfd119be143194d683
[preflight] Running pre-flight checks
        [WARNING IsDockerSystemdCheck]: detected "cgroupfs" as the Docker cgroup driver. The recommended driver is "systemd". Please follo
w the guide at https://kubernetes.io/docs/setup/cri/
        [WARNING SystemVerification]: this Docker version is not on the list of validated versions: 20.10.14. Latest validated version: 19
.03
[preflight] Reading configuration from the cluster...
[preflight] FYI: You can look at this config file with 'kubectl -n kube-system get cm kubeadm-config -oyaml'
[kubelet-start] Writing kubelet configuration to file "/var/lib/kubelet/config.yaml"
[kubelet-start] Writing kubelet environment file with flags to file "/var/lib/kubelet/kubeadm-flags.env"
[kubelet-start] Starting the kubelet
[kubelet-start] Waiting for the kubelet to perform the TLS Bootstrap...

This node has joined the cluster:
* Certificate signing request was sent to apiserver and a response was received.
* The Kubelet was informed of the new secure connection details.

Run 'kubectl get nodes' on the control-plane to see this node join the cluster.
```

图 9-18　将 node1 加入集群

第十三步，在 master 节点上查看 node 节点的状态，操作命令如下：

```
[root@master ~]# kubectl get nodes
[root@master ~]# kubectl get pods -n kube-system
```

命令运行结果如图 9-19 所示。

```
[root@master ~]# kubectl get nodes
NAME     STATUS   ROLES    AGE     VERSION
master   Ready    master   22h     v1.19.0
node1    Ready    <none>   4h17m   v1.19.0
node2    Ready    <none>   36m     v1.19.0
[root@master ~]# kubectl get pods -n kube-system
NAME                             READY   STATUS    RESTARTS   AGE
coredns-6d56c8448f-tbwfn         1/1     Running   0          22h
coredns-6d56c8448f-tmk74         1/1     Running   0          22h
etcd-master                      1/1     Running   0          22h
kube-apiserver-master            1/1     Running   0          22h
kube-controller-manager-master   1/1     Running   2          22h
kube-proxy-d5t87                 1/1     Running   0          4h17m
kube-proxy-mkrbk                 1/1     Running   0          22h
kube-proxy-qmjkf                 1/1     Running   0          37m
kube-scheduler-master            1/1     Running   2          22h
```

图 9-19　在 master 节点上查看 node 节点状态

9.2.3　Kubectl 的基本操作

Kubectl 的基本操作

1. 任务目标

掌握 Kubectl 命令的基本用法。

2. 任务内容

(1) 检查 Kubernetes 集群各节点状态。

(2) 查看 Kubernetes 版本信息、自带的资源对象和系统 Pod 状态。

(3) 创建 Pod 并查看其详细信息。

(4) 登录容器查看其中的内容。

(5) 删除 Pod。

3. 完成任务所需的设备和软件

(1) 一台安装 Windows 10 操作系统的计算机。

(2) VMware Workstation、Docker。

(3) 远程管理工具 MobaXterm。

4. 任务实施步骤

第一步，检查 Kubernetes 集群各节点状态，确保各节点状态正常，操作命令如下：

[root@master ~]# kubectl get nodes

命令运行结果如图 9-20 所示。

```
[root@master ~]# kubectl get nodes
NAME     STATUS   ROLES    AGE    VERSION
master   Ready    master   251d   v1.19.0
node1    Ready    <none>   251d   v1.19.0
node2    Ready    <none>   251d   v1.19.0
[root@master ~]#
```

图 9-20 检查 Kubernetes 集群各节点状态

第二步，查看 Kubernetes 版本信息，操作命令如下：

[root@master ~]# kubectl version --short

命令运行结果如图 9-21 所示。

```
[root@master ~]# kubectl version --short
Client Version: v1.19.0
Server Version: v1.19.0
[root@master ~]#
```

图 9-21 检查 Kubernetes 版本信息

第三步，查看 Kubernetes 自带的资源对象，操作命名如下：

[root@master ~]# kubectl api-resources

命令运行结果如图 9-22 所示。

```
[root@master ~]# kubectl api-resources
NAME                              SHORTNAMES   APIGROUP                       NAMESPACED   KIND
bindings                                                                      true         Binding
componentstatuses                 cs                                          false        ComponentStatus
configmaps                        cm                                          true         ConfigMap
endpoints                         ep                                          true         Endpoints
events                            ev                                          true         Event
limitranges                       limits                                      true         LimitRange
namespaces                        ns                                          false        Namespace
nodes                             no                                          false        Node
persistentvolumeclaims            pvc                                         true         PersistentVolumeClaim
persistentvolumes                 pv                                          false        PersistentVolume
pods                              po                                          true         Pod
podtemplates                                                                  true         PodTemplate
replicationcontrollers            rc                                          true         ReplicationController
resourcequotas                    quota                                       true         ResourceQuota
secrets                                                                       true         Secret
serviceaccounts                   sa                                          true         ServiceAccount
services                          svc                                         true         Service
mutatingwebhookconfigurations                  admissionregistration.k8s.io   false        MutatingWebhookConfiguration
validatingwebhookconfigurations                admissionregistration.k8s.io   false        ValidatingWebhookConfiguration
customresourcedefinitions         crd,crds     apiextensions.k8s.io           false        CustomResourceDefinition
apiservices                                    apiregistration.k8s.io         false        APIService
controllerrevisions                            apps                           true         ControllerRevision
daemonsets                        ds           apps                           true         DaemonSet
deployments                       deploy       apps                           true         Deployment
replicasets                       rs           apps                           true         ReplicaSet
statefulsets                      sts          apps                           true         StatefulSet
tokenreviews                                   authentication.k8s.io          false        TokenReview
localsubjectaccessreviews                      authorization.k8s.io           true         LocalSubjectAccessReview
selfsubjectaccessreviews                       authorization.k8s.io           false        SelfSubjectAccessReview
selfsubjectrulesreviews                        authorization.k8s.io           false        SelfSubjectRulesReview
subjectaccessreviews                           authorization.k8s.io           false        SubjectAccessReview
horizontalpodautoscalers          hpa          autoscaling                    true         HorizontalPodAutoscaler
cronjobs                          cj           batch                          true         CronJob
jobs                                           batch                          true         Job
certificatesigningrequests        csr          certificates.k8s.io            false        CertificateSigningRequest
leases                                         coordination.k8s.io            true         Lease
endpointslices                                 discovery.k8s.io               true         EndpointSlice
```

图 9-22 查看 Kubernetes 自带的资源对象

第四步，查看 Kubernetes 集群的系统 Pod 状态，确保所有系统 Pod 运行正常，操作命令如下：

```
[root@master ~]# kubectl get pods -n kube-system
```

命令运行结果如图 9-23 所示。

```
[root@master ~]# kubectl get pods -n kube-system
NAME                               READY   STATUS    RESTARTS   AGE
coredns-6d56c8448f-tbwfn           1/1     Running   10         251d
coredns-6d56c8448f-tmk74           1/1     Running   9          251d
etcd-master                        1/1     Running   9          251d
kube-apiserver-master              1/1     Running   10         251d
kube-controller-manager-master     1/1     Running   2          251d
kube-proxy-d5t87                   1/1     Running   5          251d
kube-proxy-mkrbk                   1/1     Running   9          251d
kube-proxy-qmjkf                   1/1     Running   4          251d
kube-scheduler-master              1/1     Running   2          251d
[root@master ~]#
```

图 9-23　查看 Kubernetes 集群的系统 Pod 状态

第五步，创建一个名称为 httpde 的 Pod，并查看 Pod 创建情况，镜像为 centos/httpd:latest，操作命令如下：

```
[root@master ~]# kubectl run http --image=centos/httpd:latest
[root@master ~]# kubectl get pod
```

命令运行结果如图 9-24 所示。

```
[root@master ~]# kubectl run http --image=centos/httpd:latest
pod/http created
[root@master ~]# kubectl get pod
NAME   READY   STATUS    RESTARTS   AGE
http   1/1     Running   0          50s
[root@master ~]#
```

图 9-24　创建一个名称为 http 的 Pod 并查看

第六步，查看 Pod 运行的节点及 IP 地址，操作命令如下：

```
[root@master ~]# kubectl get pod http -o wide
```

命令运行结果如图 9-25 所示。

```
[root@master ~]# kubectl get pod http -o wide
NAME   READY   STATUS    RESTARTS   AGE   IP            NODE    NOMINATED NODE   READINESS GATES
http   1/1     Running   0          90s   10.244.2.28   node2   <none>           <none>
[root@master ~]#
```

图 9-25　查看 Pod 运行的节点及 IP 地址

第七步，查看 Pod 详细信息，操作命令如下：

```
[root@master ~]# kubectl describe pod http
```

命令运行结果如图 9-26 所示。

```
[root@master ~]# kubectl describe pod http
Name:         http
Namespace:    default
Priority:     0
Node:         node2/192.168.1.22
Start Time:   Tue, 27 Aug 2024 06:56:46 +0800
Labels:       run=http
Annotations:  <none>
Status:       Running
IP:           10.244.2.13
IPs:
  IP:  10.244.2.13
```

```
Containers:
  http:
    Container ID:   docker://a66516040dbb5d3ae64b81a95c0c5a617a2892d18ce6f2248f4eeeba8b338f38
    Image:          centos/httpd:latest
    Image ID:       docker-pullable://centos/httpd@sha256:26c6674463ff3b8529874b17f8bb55d21a0dcf86e025eafb3c9eeee15ee4f369
    Port:           <none>
    Host Port:      <none>
    State:          Running
      Started:      Tue, 27 Aug 2024 06:56:52 +0800
    Ready:          True
    Restart Count:  0
    Environment:    <none>
    Mounts:
      /var/run/secrets/kubernetes.io/serviceaccount from default-token-crvmh (ro)
Conditions:
  Type              Status
  Initialized       True
  Ready             True
  ContainersReady   True
  PodScheduled      True
Volumes:
  default-token-crvmh:
    Type:        Secret (a volume populated by a Secret)
    SecretName:  default-token-crvmh
    Optional:    false
QoS Class:       BestEffort
Node-Selectors:  <none>
Tolerations:     node.kubernetes.io/not-ready:NoExecute op=Exists for 300s
                 node.kubernetes.io/unreachable:NoExecute op=Exists for 300s
Events:
  Type    Reason     Age   From             Message
  ----    ------     ----  ----             -------
  Normal  Scheduled  11s                    Successfully assigned default/http to node2
  Normal  Pulling    10s   kubelet, node2   Pulling image "centos/httpd:latest"
  Normal  Pulled     7s    kubelet, node2   Successfully pulled image "centos/httpd:latest" in 3.573688319s
  Normal  Created    6s    kubelet, node2   Created container http
  Normal  Started    6s    kubelet, node2   Started container http
[root@master ~]# kubectl get pod
```

图 9-26　查看 Pod 详细信息

第八步，以 YAML 格式查看 Pod 详情，操作命令如下：

[root@master ~]# kubectl get pod http -o yaml

命令运行结果如图 9-27 所示。

```
[root@master ~]# kubectl get pod http -o yaml
apiVersion: v1
kind: Pod
metadata:
  creationTimestamp: "2024-08-26T22:56:44Z"
  labels:
    run: http
  managedFields:
  - apiVersion: v1
    fieldsType: FieldsV1
    fieldsV1:
      f:metadata:
        f:labels:
          .: {}
          f:run: {}
      f:spec:
        f:containers:
          k:{"name":"http"}:
            .: {}
            f:image: {}
            f:imagePullPolicy: {}
            f:name: {}
            f:resources: {}
            f:terminationMessagePath: {}
            f:terminationMessagePolicy: {}
        f:dnsPolicy: {}
        f:enableServiceLinks: {}
        f:restartPolicy: {}
        f:schedulerName: {}
        f:securityContext: {}
        f:terminationGracePeriodSeconds: {}
    manager: kubectl-run
    operation: Update
    time: "2024-08-26T22:56:44Z"
  - apiVersion: v1
    fieldsType: FieldsV1
    fieldsV1:
      f:status:
        f:conditions:
```

图 9-27　以 YAML 格式查看 Pod 详情

第九步，将 YAML 格式的 Pod 信息导入到 YAML 文件，并查看文件内容，操作命令如下：

```
[root@master ~]# kubectl run http -o yaml --image=centos/httpd:latest --dry-run=client > http.yaml
[root@master ~]# ls
[root@master ~]# cat http.yaml
```

命令运行结果如图 9-28 所示。

```
[root@master ~]# ls
anaconda-ks.cfg  compose_lnmp  constar.tar  consul  httpd  mysql  nginx  tomcat
[root@master ~]# kubectl run http -o yaml --image=centos/httpd:latest --dry-run=client > http.yaml
[root@master ~]# ls
anaconda-ks.cfg  compose_lnmp  constar.tar  consul  httpd  http.yaml  mysql  nginx  tomcat
[root@master ~]# cat http.yaml
apiVersion: v1
kind: Pod
metadata:
  creationTimestamp: null
  labels:
    run: http
  name: http
spec:
  containers:
  - image: centos/httpd:latest
    name: http
    resources: {}
  dnsPolicy: ClusterFirst
  restartPolicy: Always
status: {}
[root@master ~]#
```

图 9-28　将 YAML 格式的 Pod 信息导入到 YAML 文件并查看文件内容

creationTimestamp：资源对象的创建时间。

resources：资源限制和请求，如内存和 CPU 的限制和请求。此处表示没有为容器指定任何资源限制或请求。

dnsPolicy：DNS 解析策略，ClusterFirst 表示在配置一个 Pod 时，其 DNS 解析将首先在 Kubernetes 集群内部进行，然后才会进行外部解析。

restartPolicy：Pod 中所有 Container 的重启策略，其值为 Always、OnFailure 或 Never，其含义如下：

- Always：只要 Container 退出，就重启，即使成功退出也要重启，默认为此值。
- OnFailure：如果 Container 的退出失败则重启。
- Never：Container 退出后永不重启。

Status：资源的当前状态信息，通常由 Kubernetes 系统自动管理。

第十步，登录到容器中查看其中的内容，操作命令如下：

```
[root@master ~]# kubectl exec -it http -- bin/sh
sh-4.2# ls
sh-4.2# exit
```

命令运行结果如图 9-29 所示。

```
[root@master ~]# kubectl exec -it http -- bin/sh
sh-4.2# ls
anaconda-post.log  bin  boot  dev  etc  home  lib  lib64  media  mnt  opt  proc  root  run  run-httpd.sh  sbin  srv  sys  tmp  usr  var
sh-4.2# exit
exit
[root@master ~]#
```

图 9-29　登录到容器中查看其中内容

第十一步，删除 Pod，操作命令如下：

```
[root@master ~]# kubectl get pod
[root@master ~]# kubectl delete pod http
[root@master ~]# kubectl get pod
```

命令运行结果如图 9-30 所示。

```
[root@master ~]# kubectl get pod
NAME   READY   STATUS    RESTARTS   AGE
http   1/1     Running   0          164m
[root@master ~]# kubectl delete pod http
pod "http" deleted
[root@master ~]# kubectl get pod
No resources found in default namespace.
[root@master ~]# _
```

图 9-30　删除 Pod

9.2.4　通过 YAML 文件创建 Pod

通过 YAML
文件创建 Pod

1. 任务目标

掌握通过 YAML 文件创建 Pod 的方法。

2. 任务内容

(1) 编写 YAML 文件。

(2) 创建 Pod。

(3) 查看 Pod 信息。

3. 完成任务所需的设备和软件

(1) 一台安装 Windows 10 操作系统的计算机。

(2) VMware Workstation、Docker。

(3) 远程管理工具 MobaXterm。

4. 任务实施步骤

第一步，参照 9.2.3 中 http.yaml 文件，编写 mynginx.yaml 文件，操作命令及代码如下：

```
[root@master ~]# vim mynginx.yaml
apiVersion: v1
kind: Pod
metadata:
  labels:
    app: mynginx
  name: mynginx
spec:
  containers:
  - image: nginx:latest
    imagePullPolicy: IfNotPresent
    name: nginx
    ports:
    - name: nginx
      protocol: TCP
      containerPort: 80
      hostPort: 30000
    resources: {}
  dnsPolicy: ClusterFirst
```

```
    restartPolicy: Never
```

第二步，通过 mynginx.yaml 创建 Pod，操作命令如下：

```
[root@master ~]# kubectl apply -f mynginx.yaml
[root@master ~]# kubectl get pod
```

命令运行结果如图 9-31 所示。

```
[root@master ~]# vim mynginx.yaml
[root@master ~]# kubectl apply -f mynginx.yaml
pod/mynginx created
[root@master ~]# kubectl get pod
NAME      READY   STATUS    RESTARTS   AGE
mynginx   1/1     Running   0          11s
```

图 9-31　通过 mynginx.yaml 创建 Pod

第三步，查看名称为 mynginx 的 Pod 详细信息，操作命令如下：

```
[root@master ~]# kubectl describe pod mynginx
```

命令运行结果如图 9-32 所示。

```
[root@master ~]# kubectl describe pod mynginx
Name:         mynginx
Namespace:    default
Priority:     0
Node:         node2/192.168.1.22
Start Time:   Wed, 11 Sep 2024 21:26:06 +0800
Labels:       app=mynginx
Annotations:  <none>
Status:       Running
IP:           10.244.2.16
IPs:
  IP:  10.244.2.16
Containers:
  nginx:
    Container ID:   docker://9d1e4391594038ac6f81f2a8e68a537bbf0bc953540f07621d5acb5b5e449a57
    Image:          nginx:latest
    Image ID:       docker://sha256:c20060033e06f882b0fbe2db7d974d72e0887a3be5e554efdb0dcf8d53512647
    Port:           80/TCP
    Host Port:      30000/TCP
    State:          Running
      Started:      Wed, 11 Sep 2024 21:26:09 +0800
    Ready:          True
    Restart Count:  0
    Environment:    <none>
    Mounts:
      /var/run/secrets/kubernetes.io/serviceaccount from default-token-crvmh (ro)
Conditions:
  Type              Status
  Initialized       True
  Ready             True
  ContainersReady   True
  PodScheduled      True
Volumes:
  default-token-crvmh:
    Type:        Secret (a volume populated by a Secret)
    SecretName:  default-token-crvmh
    Optional:    false
QoS Class:       BestEffort
Node-Selectors:  <none>
```

图 9-32　查看名称为 mynginx 的 Pod 详细信息

从图中可以看到，名称为 mynginx 的 Pod 中运行 nginx 镜像的容器 IP 地址为 10.244.2.16，该容器运行在节点 node2 上。

第四步，登录容器并访问 nginx 应用，操作命令如下：

```
[root@master ~]# kubectl exec -it mynginx -- bin/sh# curl 10.244.2.16:80
```

命令运行结果如图 9-33 所示。

```
[root@master ~]# kubectl exec -it mynginx -- bin/sh
# curl 10.244.2.16:80
<!DOCTYPE html>
<html>
<head>
<title>Welcome to nginx!</title>
<style>
html { color-scheme: light dark; }
body { width: 35em; margin: 0 auto;
font-family: Tahoma, Verdana, Arial, sans-serif; }
</style>
</head>
<body>
<h1>Welcome to nginx!</h1>
<p>If you see this page, the nginx web server is successfully installed and
working. Further configuration is required.</p>

<p>For online documentation and support please refer to
<a href="http://nginx.org/">nginx.org</a>.<br/>
Commercial support is available at
<a href="http://nginx.com/">nginx.com</a>.</p>

<p><em>Thank you for using nginx.</em></p>
</body>
</html>
```

图 9-33　登录容器并访问 nginx 应用

第五步，在浏览器中通过地址 http://192.168.1.22:30000 访问 nginx 应用，结果如图 9-34 所示。

Welcome to nginx!

If you see this page, the nginx web server is successfully installed and working. Further configuration is required.

For online documentation and support please refer to nginx.org.
Commercial support is available at nginx.com.

Thank you for using nginx.

图 9-34　通过浏览器访问 nginx 应用

第六步，编写 twocontainers.yaml 文件，操作命令及代码如下：

```
[root@master ~]# vim twocontainers.yaml
apiVersion: v1
kind: Pod
metadata:
  labels:
    app: twocontainers
  name: twocontainers
spec:
  containers:
  - image: nginx:latest
    imagePullPolicy: IfNotPresent
    name: nginx
    ports:
    - containerPort: 80
  - image: redis:latest
```

```
      imagePullPolicy: IfNotPresent
      name: redis
      ports:
      - containerPort: 8000
  dnsPolicy: ClusterFirst
  restartPolicy: Never
```

第七步，通过 twocontainers.yaml 创建 Pod，操作命令如下：

```
[root@master ~]# kubectl apply -f twocontainers.yaml
[root@master ~]# kubectl get pod
```

命令运行结果如图 9-35 所示。

```
[root@master ~]# kubectl apply -f twocontainers.yaml
pod/twocontainers created
[root@master ~]# kubectl get pod
NAME            READY   STATUS    RESTARTS   AGE
twocontainers   2/2     Running   0          11s
[root@master ~]#
```

图 9-35　通过 twocontainer.yaml 创建 Pod

第八步，查看名称为 twocontainers 的 Pod 详细信息，操作命令如下：

```
[root@master ~]# kubectl describe pod twocontainers
```

命令运行结果如图 9-36 所示。

```
[root@master ~]# kubectl describe pod twocontainers
Name:         twocontainers
Namespace:    default
Priority:     0
Node:         node2/192.168.1.22
Start Time:   Sat, 21 Sep 2024 07:11:19 +0800
Labels:       app=twocontainers
Annotations:  <none>
Status:       Running
IP:           10.244.2.26
IPs:
  IP:  10.244.2.26
Containers:
  nginx:
    Container ID:   docker://91294e64b37cc1bb038a78d8226426882b43ab2f9fd7594670078c5a251b39aa
    Image:          nginx:latest
    Image ID:       docker://sha256:c20060033e06f882b0fbe2db7d974d72e0887a3be5e554efdb0dcf8d53512647
    Port:           80/TCP
    Host Port:      0/TCP
    State:          Running
      Started:      Sat, 21 Sep 2024 07:11:21 +0800
    Ready:          True
    Restart Count:  0
    Environment:    <none>
    Mounts:
      /var/run/secrets/kubernetes.io/serviceaccount from default-token-crvmh (ro)
  redis:
    Container ID:   docker://0db22aac0ab264204e4724d2fba32002967e367d8b9267ff7313e7db6b72a5f2
    Image:          redis:latest
    Image ID:       docker-pullable://redis@sha256:eadf354977d428e347d93046bb1a5569d701e8deb68f090215534a99dbcb23b9
    Port:           8000/TCP
    Host Port:      0/TCP
    State:          Running
      Started:      Sat, 21 Sep 2024 07:11:22 +0800
    Ready:          True
    Restart Count:  0
    Environment:    <none>
    Mounts:
      /var/run/secrets/kubernetes.io/serviceaccount from default-token-crvmh (ro)
```

图 9-36　查看名称为 twocontainer 的 Pod 详细信息

可见，名为 twocontainers 的 Pod 中的两个容器运行在 node2 节点上，端口分别为 80 和 8000。

9.2.5 通过标签调度 Pod

通过标签调度 Pod

1. 任务目标

掌握通过标签调度 Pod 的方法。

2. 任务内容

(1) 为 K8S 集群中的工作节点添加标签。
(2) 创建 Pod 并指定其标签。
(3) 通过标签进行调度。

3. 完成任务所需的设备和软件

(1) 一台安装 Windows 10 操作系统的计算机。
(2) VMware Workstation、Docker。
(3) 远程管理工具 MobaXterm。

4. 任务实施步骤

第一步，在 master 节点上查看 node 节点的标签，操作命令如下：

```
[root@master ~]# kubectl get nodes --show-labels
```

命令运行结果如图 9-37 所示。

```
[root@master ~]# kubectl get nodes --show-labels
NAME     STATUS   ROLES    AGE    VERSION   LABELS
master   Ready    master   273d   v1.19.0   beta.kubernetes.io/arch=amd64,beta.kubernetes.io/os=linux,kubernetes.io/arch=amd64,kubernetes.
io/hostname=master,kubernetes.io/os=linux,node-role.kubernetes.io/master=
node1    Ready    <none>   272d   v1.19.0   beta.kubernetes.io/arch=amd64,beta.kubernetes.io/os=linux,kubernetes.io/arch=amd64,kubernetes.
io/hostname=node1,kubernetes.io/os=linux
node2    Ready    <none>   272d   v1.19.0   beta.kubernetes.io/arch=amd64,beta.kubernetes.io/os=linux,kubernetes.io/arch=amd64,kubernetes.
io/hostname=node2,kubernetes.io/os=linux
[root@master ~]#
```

图 9-37 在 master 节点上查看 node 节点的标签

第二步，修改 node 节点的标签，操作命令如下：

```
[root@master ~]# kubectl label nodes node1 labelname=node1
[root@master ~]# kubectl label nodes node2 labelname=node2
[root@master ~]# kubectl get nodes --show-labels
```

命令运行结果如图 9-38 所示。

```
[root@master ~]# kubectl label nodes node1 labelname=node1
node/node1 labeled
[root@master ~]# kubectl label nodes node2 labelname=node2
node/node2 labeled
[root@master ~]# kubectl get nodes --show-labels
NAME     STATUS   ROLES    AGE    VERSION   LABELS
master   Ready    master   273d   v1.19.0   beta.kubernetes.io/arch=amd64,beta.kubernetes.io/os=linux,kubernetes.io/arch=amd64,kubernetes.
io/hostname=master,kubernetes.io/os=linux,node-role.kubernetes.io/master=
node1    Ready    <none>   272d   v1.19.0   beta.kubernetes.io/arch=amd64,beta.kubernetes.io/os=linux,kubernetes.io/arch=amd64,kubernetes.
io/hostname=node1,kubernetes.io/os=linux,labelname=node1
node2    Ready    <none>   272d   v1.19.0   beta.kubernetes.io/arch=amd64,beta.kubernetes.io/os=linux,kubernetes.io/arch=amd64,kubernetes.
io/hostname=node2,kubernetes.io/os=linux,labelname=node2
[root@master ~]#
```

图 9-38 修改 node 节点的标签

第三步，编写 mynginx-label.yaml 文件，操作命令及代码如下：

```
[root@master ~]# vim mynginx-label.yaml
apiVersion: v1
kind: Pod
metadata:
```

```
    labels:
      app: mynginx-label
    name: mynginx-label
spec:
    containers:
    - image: nginx:latest
      imagePullPolicy: IfNotPresent
      name: nginx
      ports:
      - name: nginx
        protocol: TCP
        containerPort: 80
        hostPort: 30000
      resources: {}
    dnsPolicy: ClusterFirst
    restartPolicy: Never
    nodeSelector:
      labelname: node1
```

第四步，通过 mynginx-label.yaml 创建 Pod，操作命令如下：

```
[root@master ~]# kubectl apply -f mynginx-label.yaml
[root@master ~]# kubectl get pod
[root@master ~]# kubectl describe pod mynginx-label
```

命令运行结果如图 9-39 所示。

```
[root@master ~]# kubectl apply -f mynginx-label.yaml
pod/mynginx-label created
[root@master ~]# kubectl get pod
NAME            READY   STATUS    RESTARTS   AGE
mynginx-label   1/1     Running   0          51s
[root@master ~]# kubectl describe pod mynginx-label
Name:         mynginx-label
Namespace:    default
Priority:     0
Node:         node1/192.168.1.21
Start Time:   Fri, 13 Sep 2024 11:02:25 +0800
Labels:       app=mynginx-label
Annotations:  <none>
Status:       Running
IP:           10.244.1.19
IPs:
  IP:  10.244.1.19
Containers:
  nginx:
    Container ID:   docker://c9c1606f2ddc5ad02c47125e86c9f1d96ce647ea916de27809ddacf1f2380ee2
    Image:          nginx:latest
    Image ID:       docker://sha256:c20060033e06f882b0fbe2db7d974d72e0887a3be5e554efdb0dcf8d53512647
    Port:           80/TCP
    Host Port:      30000/TCP
    State:          Running
      Started:      Fri, 13 Sep 2024 11:02:28 +0800
    Ready:          True
    Restart Count:  0
    Environment:    <none>
    Mounts:
      /var/run/secrets/kubernetes.io/serviceaccount from default-token-crvmh (ro)
Conditions:
  Type              Status
  Initialized       True
  Ready             True
  ContainersReady   True
  PodScheduled      True
Volumes:
  default-token-crvmh:
```

图 9-39　通过 mynginx-label.yaml 创建 Pod

从图中可以看出，该 Pod 被调度到了 node1 节点上。

第五步，删除该 Pod，操作命令如下：

```
[root@master ~]# kubectl delete pod mynginx-label
[root@master ~]# kubectl get pod
```

命令运行结果如图 9-40 所示。

```
[root@master ~]# kubectl delete pod mynginx-label
pod "mynginx-label" deleted
[root@master ~]# kubectl get pod
No resources found in default namespace.
[root@master ~]#
```

图 9-40 删除该 Pod：mynginx-label

第六步，修改 mynginx-label.yaml 文件，将 Pod 调度到 node2 节点上，操作命令及代码如下：

```
[root@master ~]# vim mynginx-label.yaml
apiVersion: v1
kind: Pod
metadata:
  labels:
    app: mynginx-label
  name: mynginx-label
spec:
  containers:
  - image: nginx:latest
    imagePullPolicy: IfNotPresent
    name: nginx
    ports:
    - name: nginx
      protocol: TCP
      containerPort: 80
      hostPort: 30000
    resources: {}
  dnsPolicy: ClusterFirst
  restartPolicy: Never
  nodeSelector:
    labelname: node2
[root@master ~]# kubectl apply -f mynginx-label.yaml
[root@master ~]# kubectl get pod
[root@master ~]# kubectl describe pod mynginx-label
```

命令运行结果如图 9-41 所示。

```
[root@master ~]# vim mynginx-label.yaml
[root@master ~]# kubectl apply -f mynginx-label.yaml
pod/mynginx-label created
[root@master ~]# kubectl get pod
NAME            READY   STATUS    RESTARTS   AGE
mynginx-label   1/1     Running   0          16s
[root@master ~]# kubectl describe pod mynginx-label
Name:         mynginx-label
Namespace:    default
Priority:     0
Node:         node2/192.168.1.22
Start Time:   Fri, 13 Sep 2024 11:21:47 +0800
Labels:       app=mynginx-label
Annotations:  <none>
Status:       Running
IP:           10.244.2.25
IPs:
  IP:  10.244.2.25
Containers:
  nginx:
    Container ID:   docker://b6f3129341c2f2348dac75b275cc49fe5e79f6fa8315964f604a6ee1d5c7fda9
    Image:          nginx:latest
    Image ID:       docker://sha256:c20060033e06f882b0fbe2db7d974d72e0887a3be5e554efdb0dcf8d53512647
    Port:           80/TCP
    Host Port:      30000/TCP
    State:          Running
      Started:      Fri, 13 Sep 2024 11:21:50 +0800
    Ready:          True
    Restart Count:  0
    Environment:    <none>
    Mounts:
      /var/run/secrets/kubernetes.io/serviceaccount from default-token-crvmh (ro)
Conditions:
  Type              Status
  Initialized       True
  Ready             True
  ContainersReady   True
  PodScheduled      True
```

图 9-41　将 Pod 调度到 node2 节点上

从图中可以看出，该 Pod 被调度到了 node2 节点上。

第七步，在浏览器中通过地址 http://192.168.1.22:30000 访问 Pod 中的应用，结果如图 9-42 所示。

图 9-42　访问 Pod 中的应用

9.2.6　通过 YAML 文件创建 Deployment

通过 YAML 文件创建 Deployment

1. 任务目标

掌握通过 YAML 文件创建 Deployment 的方法。

2. 任务内容

(1) 编辑 YAML 文件并创建 Deployment。

(2) 查看 Deployment 相关信息。

(3) 更新与回滚 Deployment。

3. 完成任务所需的设备和软件

(1) 一台安装 Windows 10 操作系统的计算机。

(2) VMware Workstation、Docker。

(3) 远程管理工具 MobaXterm。

4. 任务实施步骤

第一步，编辑 nginx-deployment.yaml 文件，操作命令和代码如下：

```
[root@master ~]# vim nginx-deployment.yaml
apiVersion: apps/v1
kind: Deployment
metadata:
  name: nginx-deployment
  labels:
    app: nginx
spec:
  replicas: 3
  selector:
    matchLabels:
      app: nginx
  template:
    metadata:
      labels:
        app: nginx
    spec:
      containers:
      - name: nginx
        image: nginx:1.21.6
        imagePullPolicy: IfNotPresent
        ports:
        - containerPort: 80
```

第二步，通过 nginx-deployment.yaml 文件创建 Deployment，操作命令如下：

```
[root@master ~]# kubectl apply -f nginx-deployment.yaml
[root@master ~]# kubectl get deployment
```

命令运行结果如图 9-43 所示。

```
[root@master ~]# vim nginx-deployment.yaml
[root@master ~]# kubectl apply -f nginx-deployment.yaml
deployment.apps/nginx-deployment created
[root@master ~]# kubectl get deployment
NAME               READY   UP-TO-DATE   AVAILABLE   AGE
nginx-deployment   3/3     3            3           6s
[root@master ~]#
```

图 9-43　通过 nginx-deployment.yaml 文件创建 Deployment

第三步，查看该 Deployment 相关的 rs 信息，操作命令如下：

```
[root@master ~]# kubectl get rs
```

命令运行结果如图 9-44 所示。

```
[root@master ~]# kubectl get rs
NAME                          DESIRED   CURRENT   READY   AGE
nginx-deployment-75b69bd684   3         3         3       25m
[root@master ~]#
```

图 9-44 查看该 Deployment 相关的 rs 信息

第四步，查看该 Deployment 相关的 pod 信息，操作命令如下：

```
[root@master ~]# kubectl get pod
```

命令运行结果如图 9-45 所示。

```
[root@master ~]# kubectl get pod
NAME                                READY   STATUS    RESTARTS   AGE
nginx-deployment-75b69bd684-hdkv7   1/1     Running   0          28m
nginx-deployment-75b69bd684-sgpqs   1/1     Running   0          28m
nginx-deployment-75b69bd684-szd7t   1/1     Running   0          28m
[root@master ~]#
```

图 9-45 查看该 Deployment 相关的 pod 信息

第五步，查看该 Deployment 相关的 Pod 容器运行在哪个节点上，操作命令如下：

```
[root@master ~]# kubectl get pods -o wide
```

命令运行结果如图 9-46 所示。

```
[root@master ~]# kubectl get pods -o wide
NAME                               READY   STATUS    RESTARTS   AGE    IP            NODE    NOMINATED NODE   READINESS GATES
nginx-deployment-85dd6f4cc-29csf   1/1     Running   0          113s   10.244.1.35   node1   <none>           <none>
nginx-deployment-85dd6f4cc-v2wwm   1/1     Running   0          113s   10.244.2.44   node2   <none>           <none>
nginx-deployment-85dd6f4cc-vpl4f   1/1     Running   0          113s   10.244.2.43   node2   <none>           <none>
[root@master ~]#
```

图 9-46 查看该 Deployment 相关的 Pod 容器运行在哪个节点上

第六步，将 nginx 镜像从 nginx:1.21.6 更新为 nginx:latest 版本，操作命令如下：

```
[root@node2 ~]# kubectl set image deployment/nginx-deployment nginx=nginx:latest --record
[root@master ~]# kubectl get deployment
[root@master ~]# kubectl describe deployment nginx-deployment
```

命令运行结果如图 9-47 所示。

```
[root@master ~]# kubectl set image deployment/nginx-deployment nginx=nginx:latest --record
deployment.apps/nginx-deployment image updated
[root@master ~]# kubectl get deployment
NAME               READY   UP-TO-DATE   AVAILABLE   AGE
nginx-deployment   3/3     3            3           8m16s
[root@master ~]# kubectl describe deployment nginx-deployment
Name:                   nginx-deployment
Namespace:              default
CreationTimestamp:      Sun, 22 Sep 2024 07:50:54 +0800
Labels:                 app=nginx
Annotations:            deployment.kubernetes.io/revision: 2
                        kubernetes.io/change-cause: kubectl set image deployment/nginx-deployment nginx=nginx:latest --record=true
Selector:               app=nginx
Replicas:               3 desired | 3 updated | 3 total | 3 available | 0 unavailable
StrategyType:           RollingUpdate
MinReadySeconds:        0
RollingUpdateStrategy:  25% max unavailable, 25% max surge
Pod Template:
  Labels:  app=nginx
  Containers:
   nginx:
    Image:        nginx:latest
    Port:         80/TCP
    Host Port:    0/TCP
    Environment:  <none>
    Mounts:       <none>
  Volumes:        <none>
Conditions:
  Type           Status  Reason
  ----           ------  ------
  Available      True    MinimumReplicasAvailable
  Progressing    True    NewReplicaSetAvailable
OldReplicaSets:  <none>
NewReplicaSet:   nginx-deployment-75b69bd684 (3/3 replicas created)
Events:
  Type    Reason             Age   From                   Message
  ----    ------             ----  ----                   -------
  Normal  ScalingReplicaSet  9m3s  deployment-controller  Scaled up replica set nginx-deployment-85dd6f4cc to 3
  Normal  ScalingReplicaSet  64s   deployment-controller  Scaled up replica set nginx-deployment-75b69bd684 to 1
```

图 9-47 将 nginx 镜像从 nginx:1.21.6 更新为 nginx:latest 版本

第七步，将 Deployment 回滚到旧版本，操作命令如下：

[root@master ~]# kubectl rollout undo deployment/nginx-deployment

[root@master ~]# kubectl get deployment

[root@master ~]# kubectl describe deployment nginx-deployment

命令操作结果如图 9-48 所示。

```
[root@master ~]# kubectl rollout undo deployment/nginx-deployment
deployment.apps/nginx-deployment rolled back
[root@master ~]# kubectl get deployment
NAME               READY   UP-TO-DATE   AVAILABLE   AGE
nginx-deployment   3/3     3            3           23m
[root@master ~]# kubectl describe deployment nginx-deployment
Name:                   nginx-deployment
Namespace:              default
CreationTimestamp:      Sun, 22 Sep 2024 07:50:54 +0800
Labels:                 app=nginx
Annotations:            deployment.kubernetes.io/revision: 3
Selector:               app=nginx
Replicas:               3 desired | 3 updated | 3 total | 3 available | 0 unavailable
StrategyType:           RollingUpdate
MinReadySeconds:        0
RollingUpdateStrategy:  25% max unavailable, 25% max surge
Pod Template:
  Labels:  app=nginx
  Containers:
   nginx:
    Image:        nginx:1.21.6
    Port:         80/TCP
    Host Port:    0/TCP
    Environment:  <none>
    Mounts:       <none>
  Volumes:        <none>
Conditions:
  Type           Status  Reason
  ----           ------  ------
  Available      True    MinimumReplicasAvailable
  Progressing    True    NewReplicaSetAvailable
OldReplicaSets:  <none>
NewReplicaSet:   nginx-deployment-85dd6f4cc (3/3 replicas created)
Events:
  Type    Reason             Age                 From                   Message
  ----    ------             ----                ----                   -------
  Normal  ScalingReplicaSet  15m                 deployment-controller  Scaled up replica set nginx-deployment-75b69bd684 to 1
  Normal  ScalingReplicaSet  15m                 deployment-controller  Scaled down replica set nginx-deployment-85dd6f4cc to 2
  Normal  ScalingReplicaSet  15m                 deployment-controller  Scaled up replica set nginx-deployment-75b69bd684 to 2
```

图 9-48　将 Deployment 回滚到旧版本

第八步，查看 Deployment 的历史版本，操作命令如下：

[root@master ~]# kubectl rollout history deployment/nginx-deployment

命令运行结果如图 9-49 所示。

```
[root@master ~]# kubectl rollout history deployment/nginx-deployment
deployment.apps/nginx-deployment
REVISION  CHANGE-CAUSE
2         kubectl set image deployment/nginx-deployment nginx=nginx:latest --record=true
3         <none>

[root@master ~]#
```

图 9-49　查看 Deployment 的历史版本

第九步，删除 Deployment，操作命令如下：

[root@master ~]# kubectl delete deployment nginx-deployment

命令运行结果如图 9-50 所示。

```
[root@master ~]# kubectl delete deployment nginx-deployment
deployment.apps "nginx-deployment" deleted
[root@master ~]#
```

图 9-50　删除 Deployment

9.2.7 多容器共享 Volume

多容器共享 Volume

1. 任务目标

了解多容器共享 Volume 的方法。

2. 任务内容

(1) 编辑 share-volume.yaml 文件并创建 Pod。
(2) 比较 busybox 容器和 tomcat 容器的日志内容。

3. 完成任务所需的设备和软件

(1) 一台安装 Windows 10 操作系统的计算机。
(2) VMware Workstation、Docker。
(3) 远程管理工具 MobaXterm。

4. 任务实施步骤

第一步，编写 Pod 定义文件，操作命令和代码如下：

```
[root@master ~]# vim share-volume.yaml
apiVersion: v1
kind: Pod
metadata:
  name: share-volume
spec:
  containers:
    - name: tomcat
      image: tomcat
      ports:
        - containerPort: 8080
      volumeMounts:
        - name: share-volume
          mountPath: /usr/local/tomcat/logs
    - name: busybox
      image: busybox
      command: ["sh","-c","tail -f /logs/catalina*.log"]
      volumeMounts:
        - name: share-volume
          mountPath: /logs
  volumes:
    - name: share-volume
      emptyDir: {}
```

第二步，通过 share-volume.yaml 创建 Pod 并查看其状态，操作命令如下：

```
[root@master ~]# kubectl create -f share-volume.yaml
```

[root@master ~]# kubectl get pod

命令运行结果如图 9-51 所示。

```
[root@master ~]# kubectl create -f share-volume.yaml
pod/share-volume created
[root@master ~]# kubectl get pod
NAME           READY   STATUS              RESTARTS   AGE
share-volume   0/2     ContainerCreating   0          5s
[root@master ~]# kubectl get pod
NAME           READY   STATUS    RESTARTS   AGE
share-volume   2/2     Running   0          111s
[root@master ~]#
```

图 9-51　通过 share-volume.yaml 创建 Pod 并查看其状态

第三步，查看 busybox 容器的日志内容，操作命令如下：

[root@master ~]# kubectl logs share-volume -c busybox

命令运行结果如图 9-52 所示。

```
[root@master ~]# kubectl logs share-volume -c busybox
22-Sep-2024 03:35:09.627 INFO [main] org.apache.catalina.startup.VersionLoggerListener.log Command line argument: -Dcatalina.home=/usr/local/tomcat
22-Sep-2024 03:35:09.627 INFO [main] org.apache.catalina.startup.VersionLoggerListener.log Command line argument: -Djava.io.tmpdir=/usr/local/tomcat/temp
22-Sep-2024 03:35:09.640 INFO [main] org.apache.catalina.core.AprLifecycleListener.lifecycleEvent Loaded Apache Tomcat Native library [2.0.8] using APR version
[1.7.2].
22-Sep-2024 03:35:09.650 INFO [main] org.apache.catalina.core.AprLifecycleListener.initializeSSL OpenSSL successfully initialized [OpenSSL 3.0.13 30 Jan 2024]
22-Sep-2024 03:35:10.672 INFO [main] org.apache.coyote.AbstractProtocol.init Initializing ProtocolHandler ["http-nio-8080"]
22-Sep-2024 03:35:10.762 INFO [main] org.apache.catalina.startup.Catalina.load Server initialization in [1948] milliseconds
22-Sep-2024 03:35:10.887 INFO [main] org.apache.catalina.core.StandardService.startInternal Starting service [Catalina]
22-Sep-2024 03:35:10.887 INFO [main] org.apache.catalina.core.StandardEngine.startInternal Starting Servlet engine: [Apache Tomcat/10.1.30]
22-Sep-2024 03:35:10.964 INFO [main] org.apache.coyote.AbstractProtocol.start Starting ProtocolHandler ["http-nio-8080"]
22-Sep-2024 03:35:11.006 INFO [main] org.apache.catalina.startup.Catalina.start Server startup in [237] milliseconds
[root@master ~]#
```

图 9-52　查看 busybox 容器的日志内容

第四步，进入到 tomcat 容器查看其日志文件，操作命令如下：

[root@master ~]# kubectl exec -it share-volume -c tomcat bash

root@share-volume:/usr/local/tomcat# ls

root@share-volume:/usr/local/tomcat# cd logs

root@share-volume:/usr/local/tomcat/logs# ls

root@share-volume:/usr/local/tomcat/logs# tail catalina.2024-09-22.log

命令运行结果如图 9-53 所示。

```
[root@master ~]# kubectl exec -it share-volume -c tomcat bash
kubectl exec [POD] [COMMAND] is DEPRECATED and will be removed in a future version. Use kubectl exec [POD] -- [COMMAND] instead.
root@share-volume:/usr/local/tomcat# ls
bin  BUILDING.txt  conf  CONTRIBUTING.md  lib  LICENSE  logs  native-jni-lib  NOTICE  README.md  RELEASE-NOTES  RUNNING.txt  temp  webapps  webapps.dist  work
root@share-volume:/usr/local/tomcat# cd logs
root@share-volume:/usr/local/tomcat/logs# ls
catalina.2024-09-22.log  localhost_access_log.2024-09-22.txt
root@share-volume:/usr/local/tomcat/logs# tail catalina.2024-09-22.log
22-Sep-2024 03:35:09.627 INFO [main] org.apache.catalina.startup.VersionLoggerListener.log Command line argument: -Dcatalina.home=/usr/local/tomcat
22-Sep-2024 03:35:09.627 INFO [main] org.apache.catalina.startup.VersionLoggerListener.log Command line argument: -Djava.io.tmpdir=/usr/local/tomcat/temp
22-Sep-2024 03:35:09.640 INFO [main] org.apache.catalina.core.AprLifecycleListener.lifecycleEvent Loaded Apache Tomcat Native library [2.0.8] using APR version
[1.7.2].
22-Sep-2024 03:35:09.650 INFO [main] org.apache.catalina.core.AprLifecycleListener.initializeSSL OpenSSL successfully initialized [OpenSSL 3.0.13 30 Jan 2024]
22-Sep-2024 03:35:10.672 INFO [main] org.apache.coyote.AbstractProtocol.init Initializing ProtocolHandler ["http-nio-8080"]
22-Sep-2024 03:35:10.762 INFO [main] org.apache.catalina.startup.Catalina.load Server initialization in [1948] milliseconds
22-Sep-2024 03:35:10.887 INFO [main] org.apache.catalina.core.StandardService.startInternal Starting service [Catalina]
22-Sep-2024 03:35:10.887 INFO [main] org.apache.catalina.core.StandardEngine.startInternal Starting Servlet engine: [Apache Tomcat/10.1.30]
22-Sep-2024 03:35:10.964 INFO [main] org.apache.coyote.AbstractProtocol.start Starting ProtocolHandler ["http-nio-8080"]
22-Sep-2024 03:35:11.006 INFO [main] org.apache.catalina.startup.Catalina.start Server startup in [237] milliseconds
root@share-volume:/usr/local/tomcat/logs#
```

图 9-53　进入到 tomcat 容器查看其日志文件

可以看出，busybox 容器的日志信息与 comcat 产生的日志信息相同。

双创视角

青云 QingCloud 在中国银行的应用

随着中国银行数据中心运维规模的不断扩大以及办公环境的日益复杂，传统架构的办公、运维环境存在数据与网络安全受威胁、信息易泄露、物理设备管理复杂、资源利用率低、高能耗等缺点。同时，对于通过多跳网络将计算和存储连接起来的传统架构来说，不断增加的虚拟机也造成了成本浪费、性能不足和管理压力的问题，同时机架空间浪费也较大。

通过集成青云 QingCloud 桌面云解决方案，中国银行总行数据中心实现了对终端桌面的统一管理，将桌面标准化、资源 / 权限控制集中化，实现了统一、受控的云桌面池与交付服务化的管理模式。与此同时，青云 QingCloud 超融合系统帮助中行实现了融合架构，即计算与存储资源的融合部署，进而整合为一个易于管理的集成系统，从而摆脱了集中存储设备性能、稳定、成本的制约，实现了模块化的高效横向扩展，并融合智能运维与调度管理技术，成为具备高扩展性、强稳定性、易运维的云平台。

通过使用青云 QingCloud 桌面云平台，中国银行总行的数据安全性得到了提升。桌面云平台集成了统一的管理与维护能力，降低了运维与管理成本，分布式集群消除了单点故障的隐患。目前，无论是运维、开发人员还是普通办公人员，均能稳定流畅地使用桌面云，实现了资源快速共享，提升了生产效率。

项 目 小 结

本项目介绍了 Kubernetes 的基本知识、体系架构、相关概念以及 Kubernetes 集群的管理，完成了部署 Kubernetes 集群各节点的系统环境和 Kubernetes 集群、Kubectl 基本操作、通过 YAML 文件创建 Pod、通过标签调度 Pod、通过 YAML 文件创建 Deployment、多容器共享 Volume 等操作任务，让读者对 Kubernetes 集群的部署和管理有了一定的认识。

习 题 测 试

一、单选题

1. Kubernetes 系统支持用户通过 (　　) 定义服务配置，用户提交配置信息后，系统会自动完成对应用容器的创建、部署、发布、伸缩、更新等操作。

A. 镜像　　　　B. 容器

C. 仓库　　　　D. 模板

2. 自我修复是指在节点产生故障时，预期的副本数量 (　　)，终止健康检查失败的容器并部署新的容器，保证服务不会中断。

A. 减少　　　　B. 增多

C. 不变　　　　D. 清零

3. Kubernetes 集群通过 (　　) 进行管理。

A. API　　　　B. Proxy

C. YAML 定义文件　　　　D. Kubectl 客户端工具

二、多选题

1. Kubernetes 遵循微服务架构理论，将整个系统划分为多个功能各异的组件，各组件(　　)，可以非常方便地运行于系统环境之中。

A. 结构清晰　　B. 结构模糊

C. 部署简单　　D. 部署复杂

2. Kubernetes 集群主要由(　　)和多个(　　)组成，两种节点上分别运行着不同的组件。

A. 仓库　　B. 容器

C. 控制节点　　D. 工作节点

3. 控制节点 Master 上部署了三种组件，它们是(　　)。

A. API Server　　B. Scheduler

C. Controller Manager　　D. Docker Swarm

三、简答题

1. 简述 Kubernetes 的主要功能。

2. 简述在 Node 节点上部署的组件的功能。

参 考 文 献

[1] 肖睿，刘震．Docker 容器技术与高可用实战 [M]．北京：人民邮电出版社，2020.

[2] 千锋教育高教产品研发部．Linux 容器云实战：Docker 与 Kubernetes 集群 (慕课版) [M]．北京：人民邮电出版社，2021.

[3] 曹政，王楠，李宝安，等．基于微服务容器化的分布式开源社区实现 [J]．软件导刊．2023，22(07)：85-97.

[4] 罗汉新，王金双．Docker 容器安全风险和防御综述 [J]．信息安全与通信保密，2022 (08)：83-93.